Climate Change, The Fourth Industrial Revolution and Public Pedagogies

Climate Change, The Fourth Industrial Revolution and Public Pedagogies: The Case for Ecosocialism uses public pedagogy as a theoretical lens to examine climate change emergency and presents a solution to the issue in ecosocialism.

The book addresses the climate's relationship with capitalism and the role of activism in highlighting the climate change emergency. With respect to the Fourth Industrial Revolution, Cole assesses the pro-capitalist arguments that this revolution can be considered a progressive force and critiques them from a Marxist perspective. A case is made for ecosocialism, a form of socialism that is informed by feminism, inclusivity and real democracy. Ecosocialism, it is argued, can address climate change destruction and harness the potential fruits of the Fourth Industrial Revolution for the good of all. The book ends by addressing the other great threat to civilisation alongside climate change, with a postscript providing some final words of warning about the dual perils of climate change and nuclear warfare.

This highly topical book will be of interest to scholars, postgraduate students and researchers, as well as to advanced undergraduate students in the fields of environmental studies, pedagogy, and sociology. It will also appeal to all readers who are concerned with the onward march of climate change destruction.

Mike Cole is Professor in Education in The International Centre for Public Pedagogy at the University of East London, UK. He is the author of *Trump, the Alt-Right and Public Pedagogies of Hate and for Fascism: What is To Be Done?* (Routledge, 2020) and *Theresa May, The Hostile Environment and Public Pedagogies of Hate and Threat: The Case for a Future Without Borders* (Routledge, 2020).

Climate Change, The Fourth Industrial Revolution and Public Pedagogies

The Case for Ecosocialism

Mike Cole

LONDON AND NEW YORK

First published 2021
byRoutledge
2 Park Square, Milton Park, Abingdon, Oxon OX14 4RN

and by Routledge
605 Third Avenue, New York, NY 10017

Routledge is an imprint of the Taylor & Francis Group, an informa business

Copyright © 2021 Mike Cole

British Library Cataloguing-in-Publication Data
A catalogue record for this book is available from the British Library

Library of Congress Cataloging-in-Publication Data
A catalog record for this book has been requested

ISBN 13: 978-0-367-50817-3 (hbk)
ISBN 13: 978-0-367-50818-0 (pbk)
ISBN 13: 978-1-003-05141-1 (ebk)

Typeset in Times New Roman
by Apex CoVantage, LLC

Contents

Acknowledgements

My grateful thanks are due to the following for help, advice and comments in the course of writing this book: Rachel Adams; Lyka Cole; Rita Cole; Dave Hill; Richard Kahn; Vinnie Kidd; Kaori Kitagawa; Peter Mayo; Peter McLaren; and Glenn Rikowski. Special thanks to John Preston for his ongoing assistance and theoretical insights into 'preparedness pedagogy'. I take full responsibility for any shortcomings in the analysis.

Introduction

It could be argued that two topics dominate world politics and economics in the second decade of the twenty-first century. These are the onward march of climate change destruction and how to stop it, and the relentless pace of technological change and how to regulate it.[1] In this book, the case is made that the only viable answer to both questions and the solution to impending catastrophe entails a 'movement of movements', encompassing a scientifically informed and serious commitment to lasting ecological sustainability in tandem with a twenty-first century vision of socialism in pursuit of 'a great transition' towards ecosocialism (Löwy, 2018),[2] an ecosocialism that is ecofeminist (e.g. Brownhill and Turner, 2020). Specifically, the book uses public pedagogy (put simply, educational activity and learning that occurs outside of formal educational institutions in the sense of schools, colleges and universities) as a theoretical lens through which to analyse discourses around climate change and the Fourth Industrial Revolution (4IR), and interrogates these with Marxist and ecosocialist theory and practice (praxis). I begin by outlining the concept of public pedagogy.

Public pedagogy

Promoting progressive social change

Social justice educator Roger Simon (1995, 109) has argued that pedagogy as a concept lends itself to a variety of sites for education to take place, that are 'multiple, shifting and overlapping'. *Public* pedagogy extends pedagogical analysis beyond schools, colleges and universities to learning in other institutions such as museums, zoos and libraries, as well as informal educational sites like popular culture, commercial spaces and the media, including of course, social media. It also occurs through figures and sites of activism, including public intellectuals and grassroots social movements (Sandlin et al., 2010).[3] It is also a widely used medium in the speeches, tweets and interviews of politicians and other public figures, and in podcasts and video lectures as well as blogs, articles and books. In the Fourth Industrial Revolution, public pedagogy is increasing exponentially. It can be regressive or progressive. With respect to progressive social and socialist and ecosocialist movements and parties, public pedagogy is definitively not intended to replace more traditional struggles in the workplaces, on the streets and in communities but to complement them.

Public Pedagogy is an important corrective to the common parlance notion that pedagogy only takes place in schools, colleges and universities.

While the parameters of the concept to public pedagogy are wide-ranging, traditionally the overwhelming focus of the majority of historical and contemporary public pedagogy theorists has been on the promotion of social justice for all. To this end, as Sandlin et al. (2011) point out, many have been involved in a counter-hegemonic project against neoliberalism and its multiple manifestations per se, and/or against the oppression of multiple identities based on gender, 'race', age, sexual orientation, and social class that it upholds. Moreover:

> Although the context and meaning of . . . [public pedagogy] differ in early sources from current parlance, in some ways . . . [it] remains consistent – the term . . . [dating back to 1894] implied a form of educational discourse in the service of the *public good.*
>
> (Sandlin et al., 2011, 341–342)

In a recent book on Donald J. Trump and the alt-right (Cole, 2019a), I introduce public pedagogies that address contemporary American political realities, including anti-fascist (94–95), anti-capitalist and pro-socialist formulations (97–115), thus going beyond the social justice agenda of progressive public pedagogy theory, expanding it to include not just ongoing struggles against the growing threat of neo-nazism – but the challenge to the capitalist system itself and the promotion of the socialist alternative. I advocate Marxism as a theory that, unlike many other progressive theories, provides both a rigorous critique of capitalism and an emblematic vision for the future (this is developed in detail in Cole, 2008, 2018). In addition, I discuss a 'public pedagogy for ecology' (Cole, 2019a, 111) and a 'public pedagogy of love' (Cole, 2019a, 101–102).

In a parallel volume (Cole, 2020b) that focuses on ex-UK Prime Minister Theresa May and the way in which racism was exacerbated via her creation of the 'really hostile environment' during her periods as Home Secretary and Prime Minister, I continue to address socialist public pedagogy, this time primarily in the UK, making the case for a socialist future without borders, viewing a borderless future as inevitable as well as socially just.

To conclude this section of the Introduction, it should be stressed that public pedagogy against capitalism and for socialism is not, of course, new, and dates back at least to the early socialists (see Cole, 2008, 13–27). One of the most auspicious and powerful public pedagogical treatises, Marx and Engels' (1848) *Manifesto of the Communist Party*, has sold around 500 million copies, and is one of the four best-selling books of all time. Both the *Manifesto* and Marx's three volumes of *Capital* (1887, 1893, 1894) are UNESCO World Heritage documents (deutschland.de., 2018). However, given that Sandlin et al. (2011) limited their analysis of 'public pedagogy' to 'public pedagogy literature', that is to say to scholarly works that actually use the term,[4] a discussion of Marx or Engels or socialism was not present in that literature at that time and was therefore not addressed.[5] It is worth noting here that during 2019, the World Socialist Web Site (WSWS), a formidable organ of public pedagogy, experienced an enormous growth in its readership.[6] The total number of page views increased to 20 million, from 14 million in 2018 (a growth of more than 40 percent). The largest period of readership, with more than two million people accessing the site each month, corresponded with the General Motors strike and the auto workers' struggle in September and October in the US (North and Kishore, 2020).

Promoting regressive social change

Public pedagogy analysis has also been deployed to look at ways in which regressive (or reactionary) discourses and accompanying policies are permeated. Thus Henry Giroux (2010, 7) refers to a 'public pedagogy of hate' in the US, emitted by a 'right-wing spin machine', influenced by the right-wing media, in particular conservative radio talk show hosts, that 'endlessly spews out a toxic rhetoric' against Muslims, African Americans and other people of colour, immigrants, and many other groups (Giroux, 2010, 8).

In Cole (2019a) (chapters 2, 3 and 4; see also Cole, 2020c), I develop and extend Giroux's public pedagogy of hate to analyse how Donald Trump promotes hatred through his speeches and via Twitter. Trump's public pedagogy of hate serves not only as an attempt to 'educate' the public at large, often to promote racism, sexism and climate change misinformation, and to mock disability, but also to embolden and legitimize the views of individuals and groups associated with the alternative right or alt-right, and other far right groups with core fascist beliefs. Ongoing policies, I demonstrate, accompany Trump's public pedagogy. I also refer to 'public pedagogy in reverse', this formulation referring to Trump's describing as 'fake news', any news that he claims is published or broadcast by certain news outlets in order to undermine or discredit him: 'don't take any notice of them because you are being misinformed.' The reality is that it is the Trump administration that has developed, institutionalised and weaponised the concept of 'fake news' to serve and further its own right wing agenda (e.g. Agostinone-Wilson, 2020).

The alt-right, I argued in Cole, 2019a, were, spurred on by Trump, also clearly and manifestly engaged in public pedagogies of hate, including misogyny, but in addition actively promoted a public pedagogy for fascism, both in their quest for white supremacy and a white ethno-state and in terms of policy recommendations for a neo-Nazi USA that embodied some key elements of classic fascism. Just as Giroux (1998, 2000), in Sandlin et al.'s (2011, 344) words, 'draws on cultural studies literature that focuses on popular culture' to challenge hegemony, so do Trump and the alt-right, but from the perspective of the (far) right rather than from the left. Whereas, to reiterate, public pedagogy has traditionally been for *more* social justice and more equality, that of Trump the alt-right and other far-right factions is, from the viewpoint of progressives, for *less* justice and equality. Thus Trump and the alt-right can be viewed as attempting to undermine 'liberal democracy' (e.g. Shattuck et al., 2018). In Cole, 2020b (see also Cole, 2019b), I widen regressive public pedagogy formulations still further. Specifically, I discuss public pedagogies of hate and threat as key components of Theresa May's 'really hostile environment' directed at ('illegal') immigrants, a toxic policy that continues under the right-wing populist government of Boris Johnson (BBC News, 2020a; Bulman, 2019; Gayle et al., 2020; Gentleman, 2019a, 2019b, 2019c; Gentleman et al., 2020; McQue et al., 2019; Webber, 2019).

Preparedness pedagogy

There is also a substantive literature on 'preparedness pedagogy' ('civil defence education', 'emergency education' and 'disaster education'), relatively new, that can be seen as a form of regressive public pedagogy.[7] John Preston (2019, chapter 1) has spelt out the reactionary political ramifications of this form of pedagogy. He argues that it is used by capitalist states to ensure the continuity and stability of capitalism in the event of disasters and emergencies. It includes activities to maintain consumption. Examples are the rapid reopening of markets following 9/11 (developed in more detail in Chapter 1 of this book) and after the London Bridge Borough Market attack in 2017.

It also includes, Preston (2019, chapter 1) argues, the continuity of capitalist production such as its restoration in the face of existential threats, for example, preserving property rights in the event of nuclear war in the US. In this event, the money supply would need to be restored and property rights maintained by keeping property deeds in deep bunkers in order to 'restart' capitalism. It further includes enabling and facilitating disaster capitalism in the form of primitive accumulation, such as making new markets where they didn't exist before after a disaster. This occurred in the aftermath of Hurricane Katrina in attempts to marketise and privatise public schooling ('state schooling' in the English sense; e.g. Saltman, 2007). Public pedagogy is also used as a mode of 'responsibilisation', placing the responsibility for preparedness onto marginalised populations (usually by class or 'race') and acting as a mode of eliminationism where the State can absolve itself of responsibility (Preston, 2019, chapter 1).

A good example of both responsibilisation and eliminationism is the Grenfell Tower disaster where the instruction to 'stay put' had a devastating effect on the victims (mainly poor people of colour) and placed blame for this policy on the Fire Service rather than considering purposive austerity by the Coalition and Tory governments, and the way in which the violent gentrification of Grenfell benefited property speculators (finance capital) (Preston, 2019, chapter 1). As Preston explains, the whole area of tower block fires is under-researched and under-funded. Grenfell did not have adequate protection measures for compartmentalisation (which the fire service had been trained on) and safety was completely inadequate. The fire service were victims of austerity and did not have the resources or training, so should not be blamed for the tragedy.

Following John Holloway (e.g. Holloway, 2010), Preston argues, that since capitalism is crisis (capitalism needs workers' labour power to survive and the labour/capital relationship is always antagonistic so crisis is fundamental to capitalist production by definition[8] and pathologically produces other crises (notably environmental, as this book demonstrates), disasters increase whilst preparedness and emergency response are starved of resources. At the same time, the rich are adopting their own form of private preparedness (Billionaire Bunkers – Wainwright, 2012) – the biggest UK market for private residential bunkers is Kensington and Chelsea, the site of Grenfell Tower (Preston, 2019, chapter 1).

Is 'public pedagogy' too vague and all-encompassing to be of theoretical value?

The extensive review of the literature provided in Sandlin et al. (2011) has led those authors to question, following Glenn Savage (2010, 103), whether the wide-ranging applications of the concept of 'public pedagogy', used in mythologizing and 'totalizing' ways, has diminished 'its usefulness as a sensitizing concept for researchers interested in learning and education outside of schools' (Sandlin et al., 2011, 358), adding that throughout their review, the meaning of 'public pedagogy' 'became increasingly vague because of a general lack of clarity among many authors regarding how they are theorizing the term' (Sandlin et al., 2011, 359). Given that their review referred to the state of play a decade ago, and, as demonstrated in this Introduction, the literature has expanded even more since then, is there even more of a case to be made that the concept has become even more overloaded, a possibility that has occurred to me in my own writing on public pedagogy?

Perhaps the best response to this question is provided by Italian neo-Marxist, Antonio Gramsci, a quote from whom is cited by Sandlin et al. (2011, 343): 'every relationship of "hegemony" is an educational one' (Gramsci, 1971, 350). Underlining the universal significance of public pedagogy and its importance in class struggle, what Gramsci meant by this was that educational relationships:

> should not be restricted to the field of the strictly 'scholastic' relationships by means of which the new generation comes into contact with the old and absorbs its experiences and its historically necessary values and 'matures' and develops a personality of its own which is historically and culturally superior. This form of relationship exists throughout society as a whole and for every individual relative to other individuals. It exists between intellectual and non-intellectual sections of the population, between the rulers and the ruled, *élites* and their followers, leaders . . . and led.
>
> (Gramsci, 1971, 350)

Gramsci's concrete explication distances his version of public pedagogy from mythologizing, while from a non-relativist, anti-postmodern Marxist perspective, 'totalizing' is to be welcomed. As Peter McLaren (personal correspondence, 2007, cited in Cole, 2008, 70) once said, 'Always totalize'. This is not to deny diversity and subjectivities (present-day socialists are cognizant of and committed to eradicating *all* forms of oppression – see Chapter 3 of this book) but to insist that there is a fundamental and prerequisite duality in capitalist society based on social class that can only be resolved by the replacement of capitalism with (eco-) socialism.

It is also important to stress that Gramsci did not interpret 'intellectual' in any genetic or inherited sense. On the contrary, he believed that we are all intellectuals, that everyone has intellectual and rational faculties, but that in capitalist societies not everyone has the social function of being an intellectual (Gramsci, 1971, 9). This has important implications for public pedagogy's progressive potential. Schooling, or in French neo-Marxist Louis Althusser's terminology, the Educational Ideological State Apparatus (ISA), serves to reproduce the class structure of capitalist society, albeit massively contested, in various ways. However, 'outside of schools' public pedagogy, particularly *focused* anti-capitalist, ecosocialist public pedagogy can act as a counter-hegemonic force; hence the very purpose of writing this book. If we are all intellectuals, we are amenable to public pedagogy, of varying degrees of complexity and sophistication, throughout our lives from early to senior years, whatever hegemonic effect the education ISA has had on us.

This is not, of course, a one-way process, but a dialectical one that can lead to public pedagogical dialogue and the formation of counter-hegemonic organic intellectuals (those who are organically aligned with the working class) who, in turn, can contribute to the creation of further counter-hegemonic organic intellectuals, and so on and so on. In the course of developing Gramscian theory, Deirdre O'Neill and Mike Wayne (2017) have identified two groups of such intellectuals: those who are middle class and have distanced themselves from the ruling class, but do not democratise their practices and retain a stamp of elitism; and those who develop from *within* the working class, but who resist assimilation and neutralisation within the established institutions. The role of public pedagogy in the creation of such organic intellectuals, particularly the latter, is a further vindication of its relevance and importance.

Outline of the chapters

Chapter 1 is organised into two sections. In section 1.1, I discuss the relationship between capitalism and planetary destruction and in section 1.2, the role of activism in fostering climate change emergency, while underlining why declaring climate change emergency is urgent. I begin with a brief summary of the agreement made at the 2015 United Nations Climate Change Conference in Paris. I then consider, from a Marxist perspective, the relationship between capitalism and planetary destruction, focusing on the role of certain capitalist world leaders. Next I address the Paris climate change accord four years on: the Madrid 2019 Climate Change Conference (COP 25), suggesting that we may be on a trend for total planetary catastrophe. I conclude this section of the chapter with a look forward to Glasgow 2020 (COP 26), taking place five years after Paris, and suggest that we have a mountain to climb.

In the second section of the chapter, after briefly outlining climate change awareness's long history, consideration is given to Greta Thunberg, and movements inspired by her example, as well as Extinction Rebellion. While the struggle against climate change extinction is, for the overwhelming majority of humankind, self-evidently to be lauded, the case is made that a climate change *emergency* needs to be declared worldwide, citing a number of factors that account for this urgency that seriously threaten the survival of our planet. Next I consider the relationship between climate change and gender before concluding the chapter by stressing that it is not just the existence of humankind that is at threat, but also around a million other species.

I then lay out, in Chapter 2, also in two sections, the wide-ranging dimensions of what has become known as The Fourth Industrial Revolution (or 4.0 or 4IR) – the fusion of technologies that is blurring lines between the physical, the digital and the biological. In section 2.1, I briefly consider the trajectories of the First Industrial Revolution; the Second Industrial Revolution; and the Third Industrial Revolution, before moving on to the specific features of the Fourth. I next address the arguments of Klaus Schwab that 4IR is a progressive capitalist force. I go on look at Schwab's nod both to Trump's nationalist populism and to 'the "Greta Thunberg" effect', all in the space of a year.

In section 2.2 of Chapter 2, I present a Marxist critique of 4IR (more appropriately, Capitalism 4.0: 4IR *under global capitalism*). The case is made that pro-capitalist public pedagogies fail to address the reality of what is actually occurring and likely to occur in the immediate and long-term future. The explanatory power of Marxist critiques of capitalism is that they are structural and systemic and point to historical processes and trends that over-ride in significance the actions of individual capitalists and their apologists, however benign their intentions. In the course of the chapter, I consider Marx and his conception of value theory and technology that encompasses the tendency of the profit to fall. This has the potential to create ongoing and intensifying crises for capitalism as technological innovations advance. I move on to a consideration of Capitalism 4.0 and gender. I conclude the chapter with a look at Amazonization, at how Amazon micromanages, exploits and diminishes its workforce; at the role of Alexa, Amazon's virtual assistant; at the way Amazon is transforming the public into mini-entrepreneurs; and at the high-tech surveillance by the state facilitated by companies such as Amazon. I conclude Chapter 2 with a look at how Amazon workers are fighting back.

Sustainability is not just environmental balance, but also includes economic and social factors. Given the inherently predatory nature of capitalism, with its primary and ultimate driving force the accumulation of surplus value, it is, as an economic and social system, fundamentally incompatible with ecological survival. In the light of the seemingly insurmountable problems that I have identified that are caused by capitalism, I turn in Chapter 3 to ecosocialism, as an alternative to Capitalism 4.0. I begin by looking at nineteenth and twentieth century socialism, taking the Paris Commune of 1871 as an instance of the former, and the Russian Revolution of 1917, as an example of the latter. In my overview of events following 1917, I include a discussion of Stalinism; and the Soviet Union and ecology. I then turn my attention to ecosocialism in the twenty-first century. Now much more in the mainstream, I consider in detail Michael Löwy's 'great transition initiative' and the form that he believes it should take. I move on to Mary Mellor's critique of Löwy and other socialists for their neglect of gender and go on to make the case that ecosocialism must be ecofeminist, an ecofeminism that is fully inclusive and encompasses women of colour in the global south, as well as diverse women of all continents. Antiracism *per se* must, of course, be a key component of ecosocialism. By appropriating the public pedagogies of the pro-capitalist lobby that exalt the vast and very real potential gains that are heralded by the Fourth Industrial Revolution, ecosocialists can confidently make the case that lasting equality for humankind is becoming increasingly possible and achievable. Unlike many progressive theories, Marxism provides both a rigorous critique of capitalism and an emblematic vision for the future. It is of the utmost importance, however, that ecosocialist public pedagogies seriously and vigorously interrogate Marxism and socialism both historically and contemporaneously as both theory and practice (praxis), and in so doing, guard and warn against current and past distortions of the work of Marx and Marxists. I conclude the chapter with a consideration of whether we have arrived at a Gramscian moment.[9]

While the book focuses on the threats from climate change and the Fourth Industrial Revolution, and the case for an ecosocialist solution, fully informed by ecofeminism and antiracism, there is another existential threat, however, as identified, in tandem with the climate crisis, by the Bulletin of the Atomic Scientists (2020). The book ends, therefore, with a Postscript that provides some brief words of warning about the perils of nuclear warfare and the ways in which the barometer of its likelihood is rising with the new technologies available in Capitalism 4.0. These apocalyptic weapons, I point out, are dangerously and recklessly promoted by US politicians. Individually or together, nuclear war and climate change are potentially civilisation-ending, and as the Bulletin concludes, immediate action is required in the form of the focused and unrelenting attention of the entire world.

Notes

1 The substantive part of this book was completed before the World Health Organisation (WHO) declared the Coronavirus, Covid-19 to be a pandemic in March 2020, and so its long-term effects are, as yet, uncertain.
2 The genesis of the book is my Afterword (Cole, 2020a) in an edited collection on truth in the era of Trump (Agostinone-Wilson, 2020) and a keynote address I delivered at the 3rd International Conference on Critical Pedagogy' at the South China Normal University, Guangzhou, People's Republic of China, November 16–17, 2019.

3 For an extensive analysis of the complexities and multifarious varieties of the concept of public pedagogy, see Sandlin et al. (2011).

4 Exceptions are Cremin, 1976; Ellsworth, 2005; Lacy, 1995; Schubert, 1986). Although these writers do not specifically use the term *public pedagogy*, they 'are widely cited in the public pedagogy literature and thus can be considered foundational texts' (Sandlin et al., 2011, 340).

5 There is one mention of 'Marxian', when they state that 'Marxian perspectives on culture insist that all public policy, regardless of origin, is always shaped by the economic context of its production' (Sandlin et al., 2011, 352).

6 I make extensive use of WSWS articles in this book. I have reservations, however, about the writers on the site's marginalisation of and sometimes hostility towards struggles and issues not directly related to social class. Neither do I agree totally with the WSWS's blanket rejection of the official trade union movement, although I agree that many official and/or full-time union leaders are not to be trusted. Finally, I would not concur with its wholesale denunciation of other left parties and organisations as 'pseudo-left'.

7 For an analysis, making the case that 'preparedness pedagogy' *should* be encompassed within 'public pedagogy', and that the conceptual links between the two should be strengthened, see Kitagawa (2017).

8 This notion of 'crisis' is further explained in the section on the tendency of the rate of profit to fall in Chapter 2 of this book.

9 The crucial question of how the change from the present world (neoliberal) capitalist system to global ecosocialism might happen is outside the scope of this book, and is a much-needed endeavour. For an interesting discussion, see Harvey (2009).

1 Capitalism and planetary destruction

Activism for climate change emergency

Introduction

This chapter is organised into two sections. In section 1.1, I discuss the relationship between capitalism and planetary destruction, and in section 1.2, activism for climate change emergency. I begin section 1.1 with a brief summary of the agreement made at the 2015 United Nations Climate Change Conference in Paris. I then consider the relationship between capitalism and planetary destruction, focusing on the negative role of certain capitalist world leaders, namely Donald Trump, Jair Bolsonaro and Scott Morrison. Next I address the Paris climate change accord four years on: Madrid (COP 25), suggesting that we may be on a trend for total planetary catastrophe. I conclude section 1.1 with a forward look to Glasgow 2020 (COP 26), suggesting that we have a mountain to climb. In section 1.2, after briefly outlining the long history of climate change awareness, the role of activism in fostering climate change emergency is analysed, with reference to Greta Thunberg, and movements inspired by her example, as well as Extinction Rebellion. The case is made that a climate change *emergency* needs to be declared worldwide, citing a number of factors that seriously threaten the survival of our planet. I move on to a consideration of the relationship between climate change and gender, before concluding the chapter by stressing that it is not just the existence of humankind that is at threat, but also around a million other species.

1.1 Capitalism and planetary destruction

The Paris climate change accord

Climate change public pedagogy got a boost and a stimulus after the Paris climate agreement of 2015 that aimed to limit the global rise in temperature attributed to gases or emissions released from industry and agriculture. Nearly two hundred countries agreed to:

- Keep global temperatures 'well below' 2C above pre-industrial levels and 'endeavour to limit' them to 1.5C
- At some point between 2050 and 2100, limit greenhouse gases emitted by human activity to the same levels that trees, soil and oceans can absorb naturally
- Review each country's contribution to cutting emissions every five years
- Enable rich countries to help poorer ones by providing 'climate finance' to adapt to climate change and switch to renewable energy

(BBC News, 2019)

Such attempts to promote climate change awareness by overwhelmingly pro-capitalist governments are to be lauded, as are attempts by any other constituencies.[1] However, from a Marxist perspective a critique of the role of the world capitalist system in fermenting climate change extinction is not only necessary, but essential. The Intergovernmental Panel on Climate Change (IPCC) (2018) declared that preventing runaway global warming will require 'far-reaching transitions in energy, land . . . and industrial systems'. To even contemplate solving the unprecedented problems that we all face, as Ashley Dawson (2019) argues, 'we need a carefully planned and democratically administered emergency program for ecological reconstruction'. However, 'such a program is not remotely reconcilable with capitalism's imperatives of profit maximization and growth, not to mention private ownership of the means of production' (Dawson, 2019). In other words, as Dawson (2019) asserts: 'We need system change to beat climate change', a slogan that encouragingly for ecosocialists often appears at climate change demonstrations.

As Dawson (2019) explains, following Marxist geographer David Harvey (2010), under capitalism economies must grow at a minimum compound rate of 3 percent to remain healthy. In Harvey's words: 'Any slowdown or blockage in capital flow will produce a crisis. If our blood flow stops, then we die. If capital flow stops, then the body politic of capitalist society dies'. Harvey gives the example of 9/11:

> This simple rule was most dramatically demonstrated in the wake of the events of 9/11. Normal processes of circulation were stopped dead in and around New York City with huge ramifications for the global economy. Within five days, then Mayor Guiliani was pleading with everyone to get out their credit cards and go shopping, go to the restaurants and the Broadway shows (seats are now available!) and shortly

thereafter the President of the United States did an unprecedented thing: he appeared in a collective commercial for the airlines pleading with people to start flying again.

(Harvey, 2010)

Given the finite planetary resource, capitalism's incessant and unrelenting growth is literally killing us. A recent major report (Waheed et al., 2019) surveying hundreds of scientific studies shows a direct link over the last 50 years between economic growth, energy use and carbon emissions. Waheed et al. (2019) conclude that it is clear that 'higher energy consumption helps to boost . . . economic growth but at the cost of environmental degradation'.

Citing think tank Carbon Tracker (2013), Dawson (2019) points out that 80 percent of known fossil fuel reserves need to be kept in the ground if we are to avert temperature rise above 2C. But many of these reserves are controlled by fossil fuel corporations accountable only to investors, and to maintain their value and market share, these companies must continue to extract and sell these reserves and discover new reserves to replace them, since contraction is inimical to growth (Dawson, 2019). Free market solutions and incentives such as carbon taxes, Dawson (2019) concludes, have failed to significantly diminish fossil fuel consumption.[2]

The negative role of certain capitalist world leaders

Certain capitalist world leaders also play a major role in contributing to climate change disasters. As Somini Sengupta (2019) reported for the *New York Times*: 'This is the world we live in: Punishing heat waves, catastrophic floods, huge fires and climate conditions so uncertain that children . . . [take] to the streets en masse in global protests to demand action'.

She continues, 'But this is also the world we live in: A pantheon of world leaders who have deep ties to the industries that are the biggest sources of planet-warming emissions, are hostile to protests, or use climate science denial to score political points':

In Russia, Vladimir Putin presides over a vast, powerful petro-state. China's state-owned companies are pushing for coal projects at home and abroad, even as the country tries in other ways to tamp down emissions. Narendra Modi of India is set on expanding coal too, even as he champions solar power.

(Sengupta, 2019)

By far the biggest and most dangerous world leaders pursuing public pedagogy and accompanying actions threatening climate change extinction are the presidents of the United States of America and Brazil and the Prime Minister of Australia.

Trump

According to James Ellsmoor (2019), the US Department of Energy, under the climate-denying President of the United States, has started referring to fossil fuels as 'molecules of freedom' and specifically natural gas as 'freedom gas.' The term may have originated during a visit by US Energy Secretary Rick Perry to the European Union in April 2019, who stated:

> Seventy-five years after liberating Europe from Nazi Germany occupation, the United States is again delivering a form of freedom to the European continent. And rather than in the form of young American soldiers, it's in the form of liquefied natural gas.
>
> (cited in Ellsmoor, 2019)

In November 2019, the Trump administration formally began withdrawing from the Paris climate change agreement, the only country not signed up to it (Buncombe, 2019). Trump has repeatedly dismissed the existence of human-caused climate change, branding it as a 'hoax', while rolling back Obama-era policies aimed at tackling the crisis (Baynes, 2019a). Moreover, most disturbingly, researchers found that between 2016 and 2019 a quarter of all references to 'climate change' were removed from federal government websites. The Environmental Data and Governance Initiative (EDGI) analysed more than 5,300 pages on the websites of 23 federal agencies and found usage of the terms 'climate change,' 'clean energy', and 'adaptation' had dropped 25 percent since Trump's inauguration (Baynes, 2019b). EDGI explains the overall strategy:

> Rather than cultivating the informational resources necessary to confront climate change, the Trump administration has attempted to remove the topic from federal agency websites, a clear policy indicator in line with withdrawing from the Paris Agreement and revoking the Clean Power Plan.
>
> (cited in Baynes, 2019b)[3]

'While prominent political, journalistic, and scientific entities are sharpening the language they use to describe the climate crisis,' EDGI goes on, 'we see precisely the opposite from this administration: removal of the term "climate change" and its replacement with less clear language' (cited in

Baynes, 2019b). Chris Baynes (2019a) concludes that Trump's position puts him at odds 'with the overwhelming majority of scientists and his own government agencies, which have warned human-caused global warming is on course to have catastrophic consequences for life on Earth'.

Towards the end of 2019, Greta Thunberg was named Time magazine's Person of the Year, prompting Trump (who was hoping to get the award himself) to tell her to 'chill out' and 'work on her anger management problem', adding that she should 'go to a good old fashioned movie with a friend' (cited in Wood, 2019). He had previously responded sarcastically to her UN speech saying, 'She seems like a very happy young girl looking forward to a bright and wonderful future. So nice to see' (cited in Wood, 2019). Thunberg once replied to those whose public pedagogies of hate are directed at her: 'When haters go after your looks and differences, it means they have nowhere left to go. And then you know you're winning'. She went on, 'I have Aspergers and that means I'm sometimes a bit different from the norm. And – given the right circumstances – being different is a superpower' (cited in Wood, 2019).[4]

In January 2020, Trump attended the World Economic Forum (WEF) in Davos,[5] and was involved in what Tim Cohen (2020) describes as 'surely one of the most bizarre non-confrontational confrontations in history', with a president of the United States and a young Swedish woman going

> toe-to-toe, without mentioning each other's names, without a meeting, and without any overt acknowledgement of each others' argument. Together they symbolise the distance between climate activists and the bastions of political power.
>
> (Cohen, 2020)

In an audience that included Thunberg, Trump declared, 'We must reject the perennial prophets of doom and their predictions of the apocalypse', dismissing climate activists as fearmongering 'prophets of doom' who will cripple global economies and strip away individual liberties in what he described as a misguided mission to save the planet. He compared them to people who predicted an overpopulation crisis in the 1960s, mass starvation in the '70s, and an end of oil in the '90s:

> These alarmists always demand the same thing: absolute power to dominate, transform and control every aspect of our lives. We will never let radical socialists destroy our economy, wreck our country or eradicate our liberty.
>
> (cited in Cohen, 2020)

In a different panel, Thunberg responded: 'The facts are clear, but they are still too uncomfortable. You just leave it because you think it's too depressing and they will give up. But people will not give up. You are the ones who are giving up' (cited in Cohen, 2020). She argued that planting trees is good (Trump had promised to plant one trillion trees) but not enough; we need zero emissions (Cohen, 2020) (Just after Trump left Davos, it was revealed that BP had successfully lobbied in favour of Trump's decision to dilute a landmark environmental law, making it easier for new major infrastructure projects, such as oil pipelines and power plants, to bypass checks – Ambrose, 2020).

Thunberg was right to point out at Davos that from a 'sustainability perspective, the right, the left and the centre have all failed' and that no 'political ideology or economic structure has been able to tackle the environmental and climate emergency and create a cohesive and sustainable world,' but, as will be argued in the last chapter of this book, wrong to claim that 'it's not about politics' (cited in Cohen, 2020). That no political ideology or economic structure *has been able* to solve climate change does not mean that it cannot. In the next two chapters, I will make the case that ecosocialism is the only way that we can both save the planet and harness the technological fruits of the Fourth Industrial Revolution for the good of all.

Bolsonaro

Valdillene Urumon points out that during his election campaign, Brazil's far-right president, Jair Bolsonaro promised to divide up indigenous lands: 'That's why the ranchers voted for him. But we don't want to share our land'. She was talking to *The Guardian*'s Latin American correspondent Tom Phillips, as a fire continued to rage near her village (Phillips, 2019a). As Phillips explains, a '2,000km road and river odyssey in Brazil reveals consensus from all sides: Bolsonaro has ushered in a new age of wrecking'. Phillips is witnessing 'an inferno: a raging conflagration obliterating yet another stretch of the world's greatest rainforest . . . a catastrophic blaze . . . [in the jungle] perhaps two miles long'. Statistics produced by Brazil's own space institute – whose director was sacked in August 2019 after clashing with Bolsonaro – show the deforestation surged a Manhattan-sized area lost every day in July of that year (Phillips, 2019a).

The Guardian travelled nearly two thousand kilometres by road and river through two of the Amazon states worst affected by the fires, Rondônia and Amazonas. Along the way, indigenous leaders, wildcat goldminers, environmental activists and government officials were agreed that Bolsonaro's stripping back of protections and anti-environmental rhetoric had contributed to the scale of the fires, more than 30,000 in August 2019, alone

(Phillips, 2019a). The delight of the goldminers and those who sell to them contrasts with the growing despair of many forest dwellers whose lives were previously upended in the 1960s when Brazil's military dictatorship bulldozed roads through the Amazon (Phillips, 2019a).

Bolsonaro's actions are made all the worse by the fact that the Amazon rainforest is a crucial life-support ecosystem. Without its strength and power to generate hydrologic systems across the sky (as far north as Iowa), absorb and store carbon (CO_2), and its life-giving endless supply of oxygen, civilization would cease to exist beyond scattered tribes, here and there (Hunziker, 2019a). The Amazon is beginning to become an 'emitter of carbon', the same as coal power plants. This is not just due to wildfires and land clearing but to global warming *per se*, all of which exacerbate each other. In environmental journalist, Robert Hunziker's (2019a) words, according to the world's two leading Amazon scientists, Thomas Lovejoy and Carlos Nobre, the combination of warming temperatures, crippling wildfires, and ongoing land clearing for cattle ranching and crops 'has extended dry seasons, killed off water-sensitive vegetation and created conditions for more fire' (Hunziker, 2019a).

Unsurprisingly, like Trump, Bolsonaro has derided Thunberg. She gave her voice to growing international condemnation of a surge of anti-indigenous violence in the Amazon: 'Indigenous people are literally being murdered for trying to protect the forrest [sic] from illegal deforestation. Over and over again'. She accompanied the tweet with a video depicting the aftermath of a drive-by shooting that left two indigenous leaders dead (Phillips, 2019b). Smirking, Bolsonaro's response was 'Greta's been saying Indians have died because they were defending the Amazon. It's amazing how much space the press gives this kind of *pirralha*' (a Portuguese word that loosely translates as 'little brat' or 'pest'; Phillips, 2019b).

Brazil's Indigenous Missionary Council rights group stated in September 2019 that 153 indigenous territories had been invaded since the start of the year, noting that Bolsonaro's 'aggressive' talk was partly responsible (Phillips, 2019b).

Morrison

Brazil is not the only place where droughts have a political dimension. Writing about Australia, Kenneth Surin (2019) points out, both the centrist Labor Party and the right-wing Liberal Party 'are responsible for Australia's water crisis': as is the case in many capitalist countries with a significant agribusiness sector, 'disaster relief measures are tailored in the main to help this sector and less so small farmers'. 'The low-interest-loans scheme intended to tide-over small farmers during the crisis,' he explains, 'will not help much'.

This is because, already heavily indebted because the drought has devastated grazing lands for their animals, they will find it difficult to repay these loans (Surin, 2019). As Martin Scott (2019) elaborates, a report from the Australia Institute, one of the country's most influential public policy think tanks, has revealed the dozens of large dams have been built in Australia in recent years, but not for the public good. According to the report:

> The recently constructed dams in the Murray-Darling Basin do not help drought-stricken towns, struggling small irrigators or the wider public. They are built with taxpayer money on private land mainly for the benefit of large corporate agribusiness like Webster Limited.
>
> (cited in Scott, 2019)

The key to the government's market approach is the sale of 'water rights' that allows irrigators to be 'owed' water by virtue of a provision allowing them to take as much as 300 percent of their annual allocation in non-drought-years to 'make up' for drought years (Surin, 2019).

The neoliberal coalition between the Liberals and Nationals is headed by strident Trump supporting Liberal Prime Minister Scott Morrison (the Australian Liberal Party is a centre-right party), who in a speech in November 2019, when referring to environmental protesters, echoing Trump, mentioned a 'new breed of radical activism' that was 'apocalyptic in tone' and pledged to outlaw boycott campaigns that he argued could hurt the country's mining industry. The remarks were made to an audience at the Queensland Resources Council, an organization that represents peak mining interests in the state (Taylor A., 2019). Surin (2019) recognises the irony of Morrison's use of the term 'apocalyptic', which was also used by the media to describe the wildfires and accompanying smog that shrouded Australia's east coast. Just as Oklahoma's US Senator Jim 'Brain Freeze' Inhofe (global warming is 'the second-largest hoax ever played on the American people, after the separation of church and state' – Nicks, 2017) brought a snowball on to the Senate floor to 'prove' that global warming did not exist, Morrison once brought a lump of coal into the House Chamber to 'show' parliament they need not be afraid of coal (Surin, 2019).

On 11 November 2019, the *New York Times* reported more than 85 fires burning across Australia's east coast, with 40 out of control (Cave, 2019). Some three weeks later, *The Guardian* warned that about a third of the 146 wildfires then burning were not contained (Associated Australian Press, 2019), and later reported that some fires would take weeks to put out (Surin, 2019). In late December 2019, it was revealed that around three million hectares (7.4 million acres) of land had been burned across the country, and that nine people had been killed and more than 800 homes destroyed

(Sky News, 2019). On New Year's Day, 2020, the New South Wales Rural Fire Service reported that 916 homes had been destroyed that season, with another 363 damaged, and 8,159 saved (BBC News, 2020b). Three days later, Penrith in Greater Sydney became the hottest place on earth at 48.9C, with bushfires across Australia generating so much heat they created their own storms (Longbottom, 2020).

Early in 2020, it was reported that the bushfires turned the sky bright orange over Auckland in New Zealand, more than 1,200 miles away, with people reporting their breathing was affected by the smoke from the devastating blazes. Hours before the sun was due to set, streetlights in Auckland were turned on and motorists forced to use headlights because the plumes of smoke had made the skies so dark (Collier, 2020). A few days later skies as far as 7,000 miles away in Chile went grey in the thick smoke, with the World Meteorological Organization (WMO) citing reports that the sunset in Argentina's capital, Buenos Aires, had turned red (Binding, 2020).

In December 2019, Morrison had been criticised for going on holiday to Hawaii as the bushfire crisis worsened, with the rising public anger at his absence eventually forcing him to cut that trip short, and following mounting criticism and weeks of angry responses by those affected by the fires, Morrison conceded in January 2020 that there were 'things I could have handled on the ground much better', saying he would seek a royal commission review (BBC News, 2020c). In the same interview, however, he defended his government's approach, claiming that it took into account the effect of climate change on the bushfires. When pressed on reducing carbon emissions, he insisted the government was on track to 'meet and beat' its targets pledged under the Paris climate agreement, for Australia to cut emissions by 26 percent to 28 percent by 2030 compared to 2005 levels (BBC News, 2020c).

The following day, he insisted that people had to get used to massive bushfires and other climate-change catastrophes, and that calling out the armed forces to deal with climate-related emergencies was the 'new normal' (Head, 2020).

As Mike Head (2020), writing for the World Socialist Web Site, contends:

> Despite Morrison's evasions, his interview pointed to the underlying refusal of the ruling capitalist class to seriously address climate change because that would cut across the profit interests of the fossil fuel conglomerates and other key sections of the corporate elite.

'Reversing his many previous denials of any connection between the bushfires and climate change', Head continues Morrison said there was 'no dispute' that climate change was creating 'the longer, hotter, dryer, summer seasons'. However, he maintained that this was an unstoppable process that, over the next ten years and beyond, requires greater 'resilience' and 'adaptation' (cited in Head, 2020). Head correctly interprets his statement:

> In other words, people just have to accept the failure of the Australian government and governments around the world to halt, let alone reverse, climate change and accept the terrible consequences – higher temperatures, long and more ferocious bushfire seasons, drought and other climate-induced catastrophes.
>
> (Head, 2020)

Backing up the predictions of Australia's Bureau of Meteorology, Professor Richard Betts from the UK Met Office Hadley Centre commented on Australia as follows:

> Temperatures . . . are extreme for now but they would be normal under a world getting on for three degrees of warming, so we are seeing a sign of what would be normal conditions under a future warming world of 3 degrees.
>
> (cited in McGrath, 2020a)

Australia's Bureau of Meteorology has referred to a prolonged drought of historic proportions, while the Climate Council of Australia (CCA) alerts that by 2040 temperatures of 50C could become common in Sydney and Melbourne unless global warming is limited to 1.5C above pre-industrial levels, while long-term climate models project a continuing decline in rainfall over southern Australia for the next century (Surin, 2019).

Madrid 2019 (COP 25)[6] – the Paris climate change accord four years on: 'on a trend for total planetary catastrophe'

In total contradiction to the capitalist apologist Morrison, Professor of Climate Change Science at the University of East Anglia UK, Corinne Le Quéré has said that we can stabilise the world's climate 'if we 'bring CO2 and other long-lived greenhouse gases down to net zero emissions'. 'If we don't do it', she warns even more starkly than Betts, 'we will have much worse impacts – so what we are seeing in Australia is not the new normal, it's a transition to worse impacts' (Mcgrath, 2020a). Catastrophically, even with current government plans to limit emissions of CO2, the world is on course for around 3C of warming by the end of the twenty-first century. What then have governments been doing since the Paris Accord? According to Hunziker (2019b), as of the end of 2019, global banks had invested $1.9 trillion in fossil fuel projects since the Paris climate change accord, while global governments, including the US, China, Russia, Saudi Arabia, India, Canada, and Australia, plan to increase fossil fuels by 120 percent by 2030. In addition, between the middle of 2018 and December 2019, China added enough new coal-based power generation (43GW) to power 31 million new homes, and plans on adding another 148GW of coal-based power, which will equal the total current coal generating capacity of the European Union. Outside of its borders, China is financing 25 percent of all new worldwide coal plant construction, for example, in South Africa, Pakistan and Bangladesh. Meanwhile, in India, coal-fired power capacity increased by 74 percent between 2012 and 2019 (Hunziker, 2019b).

Few countries came to the December 2019 climate talks in Madrid (COP25) with updated plans to reach the Paris goals. Instead the focus was on narrow technical issues such as the workings of the global carbon markets, a means by which countries can trade their successes in cutting emissions with other countries that have not cut their own emissions fast enough (Harvey, 2019). According to Fiona Harvey (2019), no major breakthrough had been seriously expected at the meeting, with the US being blamed for refusing to agree to financial assistance to developing countries for the ravages they face from climate breakdown. Not unexpectedly, another wrecker was Brazil who held up agreement over a provision allowing governments to trade in carbon credits (permits allowing the holder to emit carbon dioxide or other greenhouse gases, but limiting the emission to a mass equal to one ton of carbon dioxide, the ultimate goal being to reduce the emission of greenhouse gases into the atmosphere). Its rationale was an insistence that its carbon sinks – mainly forests, including the Amazon – should count towards its emissions-cutting goals, while also selling carbon credits derived from preserving forests to other countries to count towards their emissions targets. Other countries responded that this was double counting and would undermine the carbon trading system (Harvey, 2019).The world's poorest countries, and those most vulnerable to climate chaos, came away largely disappointed and calling for more action in the next year (Harvey, 2019).

Sonam Wangdi, chair of the Least Developed Countries Group, commented:

> Our people are already suffering from the impacts of climate change. Our communities across the world are being devastated. Global emissions must be drastically and urgently reduced to limit further impacts, and financial support scaled up so our countries can better address climate change and its impacts.
>
> (cited in Harvey, 2019)

Mohamed Adow, the director of Power Shift Africa, a climate and energy think tank, added:

> This is a disastrous, profoundly distressing outcome – the worst I have ever seen. At a time when scientists are queuing up to warn about terrifying consequences if emissions keep rising, and schoolchildren taking to the streets in their millions, what we have here in Madrid is a betrayal of people across the world. It is disgraceful and governments are simply not doing their job of protecting the planet.
>
> (cited in Harvey, 2019)

Hunziker (2019b) views the world to be 'on a colossal fossil fuel growth phase in the face of stark warnings from scientists that emissions must decline to net zero'. Peter Carter, an IPCC expert reviewer, has stated:

> In COP25 the science has been blocked completely. Right now all three major greenhouse gas concentrations, carbon dioxide, methane, and nitrous oxide are accelerating. It means we are on a trend for total planetary catastrophe. We are on a trend for biosphere collapse. Carbon dioxide is on a rate exceeding anything over the past millions of years. We are at 412 ppm [parts per million]. To put that into context, we have an ice core that goes back 2.2 million years. The highest CO2 over that period is 300 ppm.
>
> (cited in Hunziker, 2019b)

Glasgow 2020 (COP 26) – the Paris climate change accord five years on: 'a mountain to climb'

Harvey (2019) concludes that the lack of progress leaves the UK, the co-host with Italy of the 2020 talks in Glasgow, also in December (the month after the US presidential election) 'with a diplomatic mountain to climb' in the coming months. Claudia Beamish (2020), MSP (Member of the Scottish Parliament) for South Scotland region and shadow cabinet secretary for climate change, environment and land reform, rightly reminds us that so much has moved forward since 2015 in terms of climate change consciousness, with Greta Thunberg lighting 'a fire in millions around the world . . . giving the young a voice and empowering them', hence COP 26 is already being lauded as the most important COP since Paris. Referring back to COP 25, she notes that it 'is disappointing to see yet again some of the richest countries in the world sidestepping their responsibilities, failing to reach agreement on finance for loss and damage, while the poorest countries in the world, who have done least to cause the crisis, face its cruellest effects already', and that 'we must go the extra mile to bring in voices from the global south and the front line of climate change', and that 'that inclusivity must also envelop activists, unionists, civil society and the voices of the many' (Beamish, 2020).

The extent of the mountainous challenge appears to be compounded by the 'huge lack of leadership and engagement' shown by Boris Johnson, according to Claire O'Neill, the sacked head of the summit. O'Neill also claims Johnson has admitted that he does not understand the issue (Mason and Harvey, 2020). After being dismissed by Johnson's Chief Special Adviser, Dominic Cummings, she sent a letter to the Prime Minister saying that the UK is 'miles off track' and that promises of action 'are not close to being met' (cited in Mason and Harvey, 2020). Moreover, she stated the cabinet sub-committee on the climate conference had not met since it formed in October 2019. 'The playground politics – the yah boo politics – has got to stop' she went on. Commenting on Johnson personally she noted:

> My advice to anybody to whom Boris is making promises – whether it is voters, world leaders, ministers, employees or indeed, to family members – is to get it in writing, get a lawyer to look at it and make sure the money is in the bank.
>
> (cited in Mason and Harvey, 2020)

Johnson hit back by pledging an earlier ban on new petrol and diesel cars (2035 rather than 2040) and called on other countries to match the UK's legal commitment to net zero carbon emisssions by 2050. However, as Rob Merrick (2020) reminds us, the car ban falls short of independent climate adviser advice of 2030, while his government faces growing criticism that it has not announced practical measures to achieve net zero.

1.2 Activism for climate change emergency

> My message is that we'll be watching you. This is all wrong. I shouldn't be up here. I should be back in school on the other side of the ocean. Yet you all come to us young people for hope. How dare you! You have stolen my dreams and my childhood with your empty words. And yet I'm one of the lucky ones. People are suffering. People are dying. Entire ecosystems are collapsing. We are in the beginning of a mass extinction, and all you can talk about is money and fairy tales of eternal economic growth. How dare you!
>
> (Greta Thunberg addresses the UN General Assembly,
> New York, September 2019)

Climate change awareness has a long history. Manuel García, Jr. (2019) alerts us to the work of Clive Thompson (2019) who has traced a CO2-induced climate change experiment back to 1856, when Eunice Newton Foote, an American suffragette, showed that carbon dioxide and water vapour were radiant-heat trapping and retaining gases, and not thermally transparent as was generally believed. In the scientific paper she submitted to the American Association for the Advancement of Science (which had to be presented by a man) she prophetically observed: 'An atmosphere of that gas would give to our earth a high temperature' (cited in García, 2019). Her results were confirmed by a number of elaborate experiments conducted by Irish physicist John Tyndall and others in subsequent years.

In the following century, during the Third Industrial Revolution (3.0), a landmark report 'Our Common Future' was published by the United Nations (United Nations, 1987), providing the definition of sustainable development as 'development that meets the needs of the present without compromising the ability of future generations to meet their own needs' (United Nations, 1987, 43). The Association for Global New Thought, whose goal is 'spiritually guided activism . . . to support concerned and informed citizens in their emerging role as ethically motivated community leaders' (Association for Global New Thought [AGNT], undated 1) has provided an overview of the report. In the light of concern about population growth, technological advance (as a result of 3.0), and consumer demand on the planetary fabric that had been smouldering away since the 1970s, a World Commission on Environment and Development was convened to address 'a new generation of environmental worries – global warming, deforestation, species loss, toxic wastes' that 'had begun to capture scientific and popular attention' (Association for Global New Thought [AGNT], undated 2).

As AGNT (undated 2) explains, the natural resources of the world were being rapidly depleted in the name of development, but the poverty that such development was supposed to solve was still widespread. By the time the report was published in 1987, population growth had ceased to be seen as a major threat, since it was mostly occurring among the planet's poor who were *not* the ones who were:

> consuming the Earth's supply of fossil fuels, warming the globe with their carbon emissions, depleting its ozone layer with their CFCs [gaseous compounds that contain carbon], poisoning soil and water with their chemicals, or wreaking ecological havoc with their oil spills. In fact, their consumption of the world's resources was minute compared to that of the industrialized world.
>
> (AGNT, undated 2)

Indeed as 'Our Common Future' declared, poverty in the developing world was less cause than effect of contemporary environmental degradation: 'the outcome of insensitive technology transfer that pauperized people and natural systems' (AGNT, undated 2); hence the necessity of sustainable development – social and economic advance to assure that humankind has a healthy and productive life – but in a form that does not compromise the ability of future generations to meet *their* needs (AGNT, undated 2). Only sustainable development can accommodate the fulfilment of human needs and the protection of air, soil, water and all forms of life – from which, ultimately, planetary stability is inseparable (AGNT, undated 2). As AGNT (undated 2) concludes, 'sustainable development brought environmentalism into poverty reduction and poverty reduction into environmentalism in a single and simple formula'. It led to the first Earth Summit – the UN Conference on Environment and Development – at Rio de Janeiro in 1992, and to the formulation of Agenda 21 (AGNT, undated 2), a comprehensive but non-binding action plan to be taken globally, nationally and locally by organizations of the United Nations System, governments, and major groups in every area in which human beings impact on the environment (United Nations, 1992).

Greta Thunberg and FridaysForFuture

Two decades after the non-binding Agenda 21, it is clear that it is climate change activism that is encouraging or forcing (by civil disobedience) a wide constituency of people to listen. A growing number of activists are prepared to engage in direct action, as well as collectively and individually taking steps to reduce carbon emissions. Ever since, at the age of 15 in August 2018, Greta Thunberg[7] began protesting outside the Swedish Parliament about the need for urgent and immediate steps to control climate change, an explosion of positive interest in the subject is omnipresent in the media, including social media (e.g. Ellis, 2019), in both publicly owned and capitalist industries worldwide, and with some exceptions as we have seen (e.g. Trump, Bolsonaro and Morrison), among most politicians. Unsurprisingly, climate deniers are backed up by reactionary think tanks and other climate denying 'institutes', funded by Exxon and Chevron (Tanuro, 2019).

FridaysForFuture (FFF) (or Youth for Climate, Climate Strike or Youth Strike for Climate) is a movement whose origins lie in Thunberg's posting on Instagram and Twitter, where it went viral, that she was continuing her protest outside the Parliament every school day for three weeks. On September 8, 2018, she decided to continue striking every Friday until Swedish Government policies provided a safe pathway well under 2C, in line with the Paris climate agreement. Hashtags #FridaysForFuture and

#Climatestrike have spread, with many students and adults began to protest outside of their parliaments and local city halls all over the world (About #FridaysForFuture, 2019).

Early in 2020, Thunberg visited Bristol in the west country of England. Here is her speech:

> This is an emergency. People are already suffering and dying from the consequences of the climate and environmental emergency but it will get worse. Still this emergency is being completely ignored by politicians, the media and those in power. Basically, nothing is being done to halt this crisis despite all the beautiful words and promises for the elected officials. So what did we do during this crucial time? What we will do right now? Well I will not stand aside and watch, I will not be silenced while the world is on fire – will you? World leaders are behaving like children, so it falls on us to be the adults in the room. Just look at Bristol as an example. The other week, the plans to expand Bristol Airport were cancelled – a lot thanks to climate activists. And of course this is far from enough, but it shows that it does actually make a difference. Activism works. So I'm telling you to act. If you look throughout history, all the great changes have come from the people. We are being betrayed by the people in power and they are failing us but we will not back down. If you feel threatened by that, then I have some very bad news for you – we will not be silenced because we are the change and change is coming whether you like it or not. Thank you and let's march.
>
> (cited in Millett, 2020)

Extinction Rebellion

Formed in London on 31 October 2018, a few months after Thunberg's intervention, and now active in large parts of the world, considerable credit must also be afforded in forcing climate change up the political agenda to Extinction Rebellion (XR) (Rehman, 2019). The purpose of its frequent series of actions in cities throughout the world is to press for action now, not by some date in the near or distant future. In so doing, it has been a most effective form of public pedagogy urging for declarations of climate emergency. As Matthew Taylor (2019) explains, the UK group, by way of example, has three core demands. First the Government must tell the truth about the scale of the crisis by declaring such an emergency and 'working with other groups and institutions to communicate the urgent need for change'. Second, the UK must drastically cut its greenhouse gas emissions, reaching net zero by 2025.[8] Third, the Government must create a citizens' assembly to hear evidence and devise policy to tackle the climate crisis. Such assemblies would bring together ordinary people to investigate, discuss and make recommendations on how to respond to the ecological emergency (Taylor M., 2019). As he points out, in the US another demand has been added:

> A just transition that prioritises the most vulnerable and indigenous sovereignty [and] establishes reparations and remediation led by and for black people, indigenous people, people of colour and poor communities for years of environmental injustice.
>
> (Taylor M., 2019)

In this context, it is important that rebellion against governments must be fully inclusive, as demanded, for example, by 'Wretched of the Earth', a diverse grassroots collective representing dozens of activist groups (see, their open letter to XR Published by the left radical UK journal, *Red Pepper*, 2019).[9]

Why declare climate change emergency?

A landmark report issued in October 2018 by The (United Nations) Intergovernmental Panel on Climate Change (IPCC), the UN's body for assessing the science related to climate change, made the case that the Paris climate agreement 2C pledge did not go far enough and that the global average temperature rise actually needs to be kept below 1.5C (Stylianou et al., 2019). *Most disturbingly, the Panel's view is that there are only about twelve years for global warming to be kept to a maximum of 1.5C: even half a degree will significantly worsen the risks of drought, floods, extreme heat and poverty for hundreds of millions of people* (Watts, 2018). Crucially, the ability to grow rice, maize and wheat will be seriously threatened with obvious potentially devastating impacts, especially for the poor in the world's developing countries, which comprise most of the southern hemisphere (World Population Review, 2020a). Almost all cities (95 percent) facing extreme climate change (rising temperatures and extreme weather) risks are in Africa or Asia, with the faster-growing cities (84 in number) most at risk (Stylianou et al., 2019).

The biggest emitter of greenhouse gases is China (26.6 percent of the total), with the US, second, at 13.1 percent (Stylianou et al., 2019). However, whereas China signed up to the Paris Accord, the US did not. If Trump gets re-elected in 2020, and his Paris agreement withdrawal comes into effect, as scheduled one day after the election on November 4, one can only speculate how this might intensify the destruction of our planet. XR's US website (Extinction Rebellion, 2019) pledges 'non-violent rebellion against the US Government for its criminal inaction on the ecological crisis.'

According to an article in *Nature* (Lenton et al., 2019), the world may have already crossed a series of 'climate tipping points'. The authors warn of 'an existential threat to civilisation', and describe 'a state of planetary emergency'. Tipping points, as environment editor for the *Guardian*, Damian Carrington (2019) explains, 'are reached when particular impacts of global heating become unstoppable, such as the runaway loss of ice sheets'. The extent of Arctic sea ice reached its lowest point on record in 2012, and the Arctic Ocean may be ice-free in the summer as soon as the 2050s unless emissions are reduced. The WMO found the extent of Arctic sea ice in 2018 was much lower than normal, with the maximum in March, the third lowest on record, and the September minimum, the sixth lowest (Stylianou et al., 2019). Extreme sea level events that used to happen once a century will occur every year in many parts of the world by 2050 because of global warming, a further IPCC panel reported in September 2019 (The Intergovernmental Panel on Climate Change) (IPCC), 2019), thus threatening the lives and livelihoods of a large percentage of the world's population, particularly in the tropics. By the middle of the century, the IPCC notes, one billion people will live in low-lying coastal areas, up from 680 million today. Even in developed countries, it will mean more coastal defences (Vaughan, 2019). This will also have an impact on food since more warming limits the mixing of the layers of water in the ocean, thus reducing the oxygen and nutrients that marine life depend on. This is already affecting fisheries and aquaculture (Nerilie Abram of the Australian National University, cited in Vaughan, 2019).

'In the past', Carrington points out, 'extreme heating of 5C was thought necessary to pass tipping points, but the latest evidence suggests this could happen between 1C and 2C' (Carrington, 2019). One tipping point, such as the release of methane from thawing permafrost (soil that is permanently frozen) may fuel others leading to a cascade (Lenton et al., 2019). Moreover, as stated by UN Secretary General António Guterres, 'the point of no return is no longer over the horizon', while the charity 'Save the Children' warns that 33 million people in Africa face emergency levels of hunger due to cyclones and droughts (McGrath, 2019).

The worst drought in a century slowed the Victoria Falls on the borders of Zimbabwe and Zambia to a trickle, with 2019 having brought 'an unprecedented decline in water levels' (Reuters, 2019). Already, southern Africa as a whole is suffering some of the worst effects of human-driven greenhouse gas emissions, with taps running dry and about 45 million people in need of food aid amid crop failures (Reuters, 2019). Richard Beilfuss, head of the International Crane Foundation, who has studied the Zambezi for decades, noted that climate change was delaying the monsoon, 'concentrating rain in bigger events, which are then much harder to store, and a much longer, excruciating dry season' (cited in Reuters, 2019).

In the northern hemisphere, the most comprehensive study into Greenland ice loss so far undertaken, involving 96 polar scientists from 50 international organisations and combining 26 separate surveys to compute changes in the mass of Greenland's ice sheet between 1992 and 2018, revealed that Greenland's ice sheet is melting seven times faster today than it was in the 1990s – and will be responsible, on current trends, for flooding 100 million people a year by the end of the century

This figure rises to 400 million people once sea level rise due to the expansion of the oceans from increasing water temperatures, as well as melting ice in the Antarctic and other regions, are also included (Bawden, 2019).

According to the World Meteorological Organization (WMO), the global average temperature for the first 10 months of 2018 was almost one degree warmer than the levels of 1850–1900 (Stylianou et al., 2019). The year 2019 did not start well. Averaged as a whole, the January 2019 global land and ocean surface temperature was 0.88°C above the twentieth-century average and tied with 2007 as the third highest temperature since global records began in 1880 (National Centers for Environmental Information, 2019). The latest available figures (January, 2020) show the highest ever temperature at 1.14°C above the twentieth-century average of 12.0°C, the fourth highest monthly temperature departure from average in the 1,681 months since 1880 (National Centers for Environmental Information, 2020).

Moreover, the 20 warmest years on record have been in the past 22 years, with 2015–2018 being the hottest four. If this trend continues, temperatures may rise by 3–5C by 2100 (Stylianou et al., 2019). In 2019, the heat in the world's oceans reached a new record that, according to Carrington (2020) shows 'irrefutable and accelerating' heating of the planet. As he explains, oceans are 'the clearest measure of the climate emergency because they absorb more than 90 percent of the heat trapped by the greenhouse gases emitted by fossil fuel burning, forest destruction and other human activities' (Carrington, 2020). Incredibly, the amount of heat being added to the oceans is the same as every person in the world running 100 microwaves all day and all night (Carrington, 2020).

A succession of heatwaves in 2018 set a number of records. As Daisy Dunne, Science writer for CarbonBrief, a UK-based website covering the latest developments in climate science, climate policy and energy policy, points out, the extreme heat 'broke temperature records simultaneously across North America, Europe and Asia' (Dunne, 2019):

> Among its impacts, the heatwave caused crop failures across Europe, fanned wildfires from Manchester in the UK to Yosemite National Park in the US and exposed more than 34,000 people to power outages in Los Angeles as the grid experienced an unprecedented demand for air conditioning.

This led scientists, she explains, to conclude that it is 'virtually certain' that the 2018 northern-hemisphere heatwave could not have happened without climate change, with "virtually certain", according to Martha Vogel, of the Institute for Atmospheric and Climate Science, amounting to more than 98 percent probability (Dunne, 2019). Moreover, summer heatwaves on such a scale could occur every year if global temperatures reach 2C above pre-industrial levels. If global warming is limited to 1.5C, such heatwaves could occur in two of every three years. By a similar token, the extreme heat seen in Japan in 2018, in which more than 1,000 people died, could not have occurred without climate change. In January 2020, NASA, NOAA (the US National Oceanic and Atmospheric Administration) and the UK Met Office all declared that the 10 years to the end of 2019 were the warmest decade on record by three global agencies (McGrath, 2020b)

Climate change and gender

Following Colleen Curry (2017), McCarthy and Sanchez (2019) argue that in the global south, women and girls are often hit hardest by climate disasters such as floods and droughts, leading to more families struggling to afford to feed and house their own children. Thus climate change is not just changing weather; its consequences are a daily reality for millions of people around the world (Qasim, 2019). As Arsheen Qasim (2019) explains, writing for ActionAid, an international charity that works with women and girls living in poverty, it is usually women and girls that struggle to survive and recover from the aftermath of climate change disasters. This is because of an already existing gender disparity: the world's richest 1 percent have more than twice the wealth as 6.9 billion people; the richest 22 men in the world have more money than all the women in Africa; women and girls do 12.5 billion hours of unpaid work per day; women's unpaid care work has a monetary value of $10.8 trillion per year (Whiting, 2020) that compounds and multiplies when disasters strike (Qasim, 2019).

From the work of ActionAid in countries like Somaliland and Bangladesh, there is evidence that climate change-related crises are 'already instigating deep, life-altering changes for some of the poorest and most marginalised women and girls in the world' (Qasim, 2019).

Increased risk of violence against women and girls

As a result of climate change-related droughts, hurricanes or cyclones, women and girls may be forced to migrate to camps for displaced people where living under temporary tarpaulin sheets or bare, plastic sheets can make them vulnerable to violence from strangers (Qasim, 2019). Qasim describes a camp for displaced people in Somaliland, where two sisters, one just 15 and the other 19 lie awake at night listening to any sounds that may alert them to men walking into their makeshift shelter which has no proper doors. Some nights, Qasim points out, they do not sleep at all for fear of being attacked.

Increased risk of child marriage

For poor families who lose their homes and livelihoods, child marriage can be seen as a way out of, or at least a way to diminish the effects of crippling poverty, while also ensuring the perceived security of their daughters from violence. Qasim gives the example of Sarmin who had to face the prospect of marriage at just 14 when floods hit Bangladesh in 2017. Her parents, having lost everything, believed that marriage would keep her well-fed and safe. As Sarmin puts it: 'Because of poverty parents marry off their daughters like me at a very early age. I cannot go to school since I got married. Life is tough for girls of young age in the village' (cited in Qasim, 2019).

More likely to miss classes or drop out of school

In times of crises, girls often have to drop out of school or miss classes because household chores become a strain with fewer family members available to share the pressures at home. Girls are often required to take care of family members, or to help with cooking, cleaning or finding water, with school considered a lower priority in times of need. Moreover, whole schools can be closed for a long period or even destroyed, while floods can make journeys to school more dangerous and longer as usual routes are destroyed (Qasim, 2019).

Increased risk of death and injury

Marginalised women and girls, the disabled and the elderly are more prone to death and injury during disasters, since as a result of their traditional roles as caretakers, women and girls often stay back in a disaster to protect their children or adults in their care while men are sometimes able to escape (Qasim, 2019). Deeply ingrained social norms can also sometimes dictate that women and girls need permission from men to leave their homes and/or are unable to escape during floods because they are not encouraged to learn to swim (Qasim, 2019).

The effect of availability of food and chances of earning a living

Many people living in pastoral or farming communities rely, of course, on the food from crops grown daily and/or reared livestock, used for milk, meat or for selling. Qasim informs us of the reality facing Nafisa from Somaliland, a twenty-year old mother, part of a pastoralist family that owned several animals that they reared for their food and income. As Qasim points out, years of climate change-related drought dried out her land and killed all her livestock. This was exacerbated when her husband took her children and left her without food or money. As she puts it:

> In many rural communities, men control the income in their households. Women who rely on men for economic support are left struggling in times of crisis when men either abandon their families, or leave to look for work elsewhere or are even killed by disasters. Women in rural communities have limited access to and ownership of their land as well which directly impacts the food they have available to eat.
>
> (Qasim, 2019)

At the same time, in Bangladesh, Shefali, 25, talks about the lack of food after the 2017 floods:

> We lost our crops. Our small goat died in the flood water. Cows are sick. My children suffered a lot during flood. Now we have no crop in our storage. I, along (with) my husband have to work in people's fields to bring food (to) our table. Life is becoming harder every year after floods.
>
> (cited in Qasim, 2019)

Loss of plant and animal species

It is not just the existence of humankind that is under threat of course. Around a million species – perhaps one eighth of all plant and animal species on Earth – are in danger of becoming extinct, many within a matter of decades, according to a report, the most comprehensive assessment of global biodiversity ever conducted, released in May, 2019 from the Intergovernmental Science-Policy Platform on Biodiversity and Ecosystem Services (IPBES). This tragic loss includes more than 40 percent of amphibians, a third of all corals and all sharks (de Vries, 2019).

A further report by the International Union for Conservation of Nature (IUCN, 2019a), the largest peer-reviewed study so far into the causes, impacts and possible solutions to ocean deoxygenation and made public in December 2019 revealed the dangers of loss of oxygen to species such as tuna, marlin and sharks (IUCN, 2019b). Ocean regions with low oxygen concentrations are expanding, with around 700 sites worldwide now affected – up from only 45 in the 1960s. In the same period, the volume of anoxic waters – areas completely depleted of oxygen – in the global ocean has quadrupled (IUCN, 2019a). Deoxygenation is changing the balance of marine life , favouring low-oxygen tolerant species (e.g. microbes, jellyfish and some squid) at the expense of low-oxygen sensitive ones (many marine species, including most fish), ultimately affecting hundreds of millions of people. Moreover, very low ocean oxygen affects basic processes like the cycling of elements crucial for life on Earth, such as nitrogen and phosphorous (IUCN, 2019b).

As Daniel de Vries (2019), writing on the World Socialist Web Site (WSWS), points out:

> While there have been five previous mass extinctions during the 3.5-billion-year history of life on Earth,[10] the die-off of biodiversity over the past 50 years is not only unprecedented in the existence of humanity, it is caused by our species.[11]

Recognising this fact, and the imminence of impending climate catastrophe, as this book goes to press (June 2020) 1501 jurisdictions in 30 countries have declared a climate emergency, amounting to over 820 million citizens (Climate Emergency Declaration, 2020).

While such declarations are essential and urgent, and while they are to be lauded, they are not enough. Referring to the IPBES report, de Vries (2019) argues that while it makes a clear call for 'transformative change,' that is, 'fundamental, system-wide reorganization across technological, economic and social factors', what is lacking is not the knowledge or technological capability to implement these changes, but the much needed social initiative. 'By its very nature', Robert Watson of IPBES observed, 'transformative change can expect opposition from those with interests vested in the status quo, but also . . . such opposition can be overcome for the broader public good' (cited in de Vries, 2019). de Vries (2019) has translated 'the cautious wording of scientific studies conducted under the auspices of the United Nations' into the language of socialism:

> the issue confronting humanity is the incapability of dealing with ecological catastrophe under the present regime: an economy based on private profit and a world divided into antagonistic nation-states. The problem is capitalism as a global system.

Capitalism is now witnessing its Fourth Industrial Revolution. This generates a whole new plethora of problems, and, having discussed the relationship between capitalism and planetary destruction in this chapter, it is to 4IR that I now turn in the next.

Notes

1 An exception is the attempt by eco-fascists to appropriate ecology (Wilson, 2019).

2 Kevin Beck (2019) has provided an explanation of exactly what is meant by 'fossil fuels'. They are produced from the remains of plants and animals that lived millions of years ago, with the 'slow transformation of this carbon-heavy material into various *hydrocarbon* compounds resulted in the creation of plentiful, highly flammable fuels'. The four types are petroleum, coal, natural gas and *Orimulsion* (italicized because it is a registered trademark name for a bitumen-based fuel). While they have a number of physical and chemical properties in common, the most critical fact about them is that they are not renewable: 'many more millions of years have to pass before even small amounts can be made again, assuming the same processes will ever even occur on the same scale' (Back, 2019). In their natural form, the carbon in fossil fuels is stored. However, when they are burned, the carbon is unlocked and returns to the atmosphere. Natural gas is considered clean-burning compared to other fossil fuels; it is getting it out of the ground that is the most problematic aspect of its production. At current levels of use, unless policies radically change, fossil fuels are expected to still account for 78 percent of energy used worldwide in 2040 (Back, 2019). Renewable energy sources ('clean' energy) include sunlight, wind, rain, tides, waves and geothermal heat (Ellabban et al., 2014).

3 The aim of the Clean Power Plan, unveiled by Barack Obama on 3 August 2015 was to cut greenhouse gas emissions from US power stations by nearly a third within 15 years, placing significant emphasis on wind, solar power and other renewable energy sources (BBC News, 2015).

4 This is not the first example of Trump's disablism (see Cole, 2019c, 18–21), not indeed sexism and hatred of women and girls (Street, 2019; see also, Cole, 2019c, 16–17). Here the misogyny of the woman-hater-in-chief (Street, 2019) is patronising, with 16 year old Thunberg demeaned as a 'young girl'.

5 The World Economic Forum and its founder are addressed in detail in the next chapter of this book on the Fourth Industrial Revolution

6 COP simply means Conference of the Parties, the supreme decision-making body for Climate Change under the United Nations Framework for the Convention on Climate Change (UNFCCC). Its task it to assess the effects of the measures taken by Parties and the progress made in achieving the ultimate objective of the Convention, on a yearly basis. The first COP meeting was held in Berlin in March 1995 (United Nations Climate Change, 2020).

7 While Thunberg is undoubtedly the most high profile female climate change activist, at least in the global north, Joe McCarthy and Erica Sanchez (2019) identify eleven others. See also Global Justice Now (2020) for six stories of women who have defended the environment and resisted corporate power

8 To underline the urgency, the UN stated in November 2019 that global emissions must fall by 7.6% every year from then until 2030 even to stay within the 1.5C ceiling and avoid 'climate chaos' (Harvey, 2019).

9 Mass action should not undermine the importance of individual and small-scale remedial action. For an interesting analysis, see Foer (2019).

10 A 'mass extinction' can be defined as a time period in which a large percentage of all known species living at the time goes extinct, or is completely wiped out (Scoville, 2018). The single biggest driver of mass extinctions appears to be major

changes in the Earth's carbon cycle, such as huge volcanoes that flooded hundreds of thousands of square miles with lava, ejecting massive amounts of heat-trapping gases such as carbon dioxide into the atmosphere, leading to global warming (Greshko and National Geographic Staff, 2019). These extinctions include the Ordovician-Silurian extinction – 444 million years ago; the Late Devonian extinction – 383–359 million years ago; the Permian-Triassic extinction – 252 million years ago; the Triassic-Jurassic extinction – 201 million years ago; and the Cretaceous-Paleogene extinction – 66 million years ago. This last extinction is the only one definitively connected to a major asteroid impact, when some 76 percent of all species on the planet, including all non-avian dinosaurs became extinct. Today, extinctions are occurring hundreds of times faster than they would naturally (for full details, see Greshko and National Geographic Staff (2019).

11   The negative effects of human impact on Earth's ecosystem has been described as the 'Anthropocene'. The popularisation of the word at the beginning of the twentieth century is credited to atmospheric chemist, Paul J. Crutzen. From a Marxist perspective, a more appropriate term might be 'Capitalocene' to refer to the destructive effects on the planet of capitalism, as discussed in this chapter. Jason W. Moore (2017, 1) stresses the need for historical thinking to be central to an understanding of capitalism's planetary crises in the twenty-first century. Arguing against the Anthropocene's 'shallow historicization', he makes the case for using the concept Capitalocene to refer to a system of power, profit and re/production in the web of life. He suggests that the longstanding environmentalist argument about the Industrial Revolution as the origin of ecological crisis ignores early capitalism's environment-making revolution (we make environments and the environments make us (Lewontin and Levins 1997, cited in Moore, 2017, 6), greater than any watershed moment since the rise of agriculture and the establishment of the first cities. While there is no question that environmental change accelerated sharply after 1850, and especially after 1945, he concludes, it seems equally pointless to explain these transformations without identifying how they fit into patterns of power, capital and nature established four hundred years earlier.

2 The Fourth Industrial Revolution (4IR)

Capitalist defence and Marxist critique

Introduction

This chapter is also in two sections. In section 2.1, I briefly consider the trajectories of the First Industrial Revolution; the Second Industrial Revolution; and the Third Industrial Revolution, before moving on to the specific features of the Fourth (4IR), the fusion of technologies that is blurring lines between the physical, the digital and the biological. I then address the argument of Klaus Schwab that 4IR is a progressive capitalist force. I go to examine Schwab's nod to both Trump's nationalist populism and to Greta Thunberg and climate change. In section 2.2, the case is made that these pro-capitalist public pedagogies fail to address the reality of what is actually occurring and likely to occur in the immediate and long-term future. I present a Marxist critique of 4IR (more appropriately, Capitalism 4.0: 4IR *under global capitalism*). The explanatory power of Marxist critiques of capitalism is that they are structural and systemic and point to historical processes and trends that over-ride in significance the seemingly progressive public pedagogies of individual supporters of capitalism and their apologists, however benign their intentions. In the course of the chapter, I consider Marx and his conception of value theory and technology that encompasses the tendency of the profit to fall. This has the potential to create ongoing and intensifying crises for capitalism as technological innovations advance. I move on to an analysis of Capitalism 4.0 and gender, arguing that the Fourth Industrial Revolution exacerbates gender inequality. I conclude the chapter with a look at Amazonization, at how Amazon micromanages, exploits and diminishes its workforce; at the role of Alexa, Amazon's virtual assistant; at the way Amazon is transforming the public into mini-entrepreneurs; and at the high-tech surveillance by the state facilitated by companies such as Amazon. I conclude Chapter 2 with a look at how Amazon workers are fighting back.

2.1   The Fourth Industrial Revolution (4IR) – capitalist defence

The four industrial revolutions

> We stand on the brink of a technological revolution that will fundamentally alter the way we live, work, and relate to one another. In its scale, scope, and complexity, the transformation will be unlike anything humankind has experienced before.

This is how academic economist, Klaus Schwab (2015; see also Schwab, 2016) (professor of business policy at the University of Geneva from 1972 to 2002) has described what he refers to as the Fourth Industrial Revolution, or 4IR (Schwab, 2018). He goes on to summarise the progression from the first to the fourth:

> The First Industrial Revolution used water and steam power to mechanize production. The Second used electric power to create mass production. The Third used electronics and information technology to automate production. Now a Fourth Industrial Revolution is building on the Third, the digital revolution that has been occurring since the middle of the last century. It is characterized by a fusion of technologies that is blurring the lines between the physical, digital, and biological spheres.
>
> (Schwab, 2015)

The First Industrial Revolution (1.0)

Desoutter Industrial (2019) has provided a useful summary of the four industrial revolutions. The First Industrial Revolution (1.0) began in the eighteenth century through the use of steam power and the mechanisation of production. For example instead of weaving looms powered by muscle, steam engines began to be used for power. Before threads were produced on simple spinning machines, whereas the mechanised version achieved eight times the volume in the same time frame. The steamship and, later, the steam-powered locomotive generated massive changes because humans and goods could move great distances in fewer hours. Steam power for industrial purposes was a major breakthrough in the constant quest to increase human productivity. Marx's co-writer, factory owner

Friedrich Engels (1872) reflected on what all this meant for the loss of worker autonomy:

> Modern industry, with its big factories and mills, where hundreds of workers supervise complicated machines driven by steam, has superseded the small workshops of the separate producers; the carriages and wagons of the highways have become substituted by railway trains, just as the small schooners and sailing feluccas have been by steamboats. . . . Let us take by way of example a cotton spinning mill. The cotton must pass through at least six successive operations before it is reduced to the state of thread, and these operations take place for the most part in different rooms. Furthermore, keeping the machines going requires an engineer to look after the steam engine, mechanics to make the current repairs, and many other labourers whose business it is to transfer the products from one room to another, and so forth. All these workers, men, women and children, are obliged to begin and finish their work at the hours fixed by the authority of the steam, which cares nothing for individual autonomy.

For Engels, the automatic machinery of the big factory was more despotic than small capitalism. Things were soon to get worse for the working class.

With respect to climate change, in the early days of the first Industrial Revolution, no one would have thought that their burning of fossil fuels would have an almost immediate effect on the climate. But a study, published in *Nature* (Abram et al., 2016), revealed that warming in some regions actually began as early as the 1830s (McGregor et al., 2016).

The Second Industrial Revolution (2.0)

About 40 years after Engels' warnings about the onslaught of industrial capitalism, the first moving assembly line was developed for the Model T Ford and began operation in 1913. Following his discovery of electricity, Thomas Edison had produced the first incandescent light bulb in 1879, while Henry Ford (1863–1947) had garnered the idea of mass production that heralded the Second Industrial Revolution (2.0) starting in the nineteenth century. Ford got the idea from a slaughterhouse in Chicago: the pigs hung from conveyor belts and each butcher performed only a part of the task of butchering the animal. Carrying this technique over into automobile production under Ford changed from one station assembling an entire automobile to vehicles being produced in partial steps on the conveyor belt. Accordingly commodities could be produced significantly faster and at lower cost (Desoutter Industrial, 2019). This increases the degree of exploitation of workers and raises profits.

As US historian Ryan Engleman (2019) concludes his article on 2.0, this Revolution fuelled the Gilded Age, a period of great extremes: 'great wealth and widespread poverty, great expansion and deep depression, new opportunities and greater standardization'. For a small minority of skilled workers, there were high wages, but for the masses economic insecurity became a basic way of life as the depressions of the 1870s and 1890s put millions out of work, while those still in work experienced extremely dangerous working conditions, with long hours, low wages, no compensation for injuries and no pensions.

The Third Industrial Revolution (3.0)

The Third Industrial Revolution (or 3.0) began in the 1970s through partial automation using memory-programmable controls and computers. The introduction of these new technologies meant that whole production processes could henceforth be automated without human assistance. Jeremy Rifkin, in a well-known analysis of 3.0 (TIR is Rifkin's acronym for 3.0), *The Zero Marginal Cost Society: the Internet of things, the collaborative commons and the eclipse of capitalism* (2014), points out that, with robot sales escalating in the US and the EU, we are witnessing the movement of 'the manufacturing sector ever closer to workless production, or what the industry calls "lights out" production'. Some examples provided by Peter Taaffe (2014), general secretary of the Socialist Party, are the music industry where, because of free internet access, pop groups and other musicians are forced to compensate by spin-offs such as gigs and memorabilia; the jobs of teachers and university lecturers threatened by the mass application of online teaching; architecture; e-book production; and the medical profession. Taaffe, however, is critical of Rifkin's conclusion that 'the capitalist era is passing', since intense competition forcing 'the introduction of ever-leaner technology' means 'the cost of actually producing each additional unit . . . become essentially zero, making the product nearly free', thus effectively drying up profit, 'the lifeblood of capitalism'. Rifkin's solution is to change society through argument and force of example. However, as Taaffe argues, capitalists 'will not calmly accept their fate'. What is required is

> not just a critique of capitalism – which Rifkin has, in a way – but also setting out in a clear way the alternative of socialism, and building the force to achieve it . . . The capitalist system cannot fully utilise the huge potential benefits flowing from the latest developments in technique.

> Only a planned economy leading to democratic socialism on a national and international scale can do this and, in the process, satisfy the yearning of those, like Jeremy Rifkin and the new generation, for real change.
>
> (Taaffe, 2014)

In Chapter 3 of this book, I discuss socialisms in the nineteenth and twentieth centuries and make the case for ecosocialism in the twenty-first, of which ecofeminism and antiracism are key components.

The Fourth Industrial Revolution (4.0)

Desoutter Industrial (2019) insists that we are currently witnessing the Fourth Industrial Revolution (or 4.0), the next step in production automation. This builds on 3.0 and is characterised by the application of information and communication technologies to industry. Production systems that already have computer technology 'are expanded by a network connection and have a digital twin on the Internet', thus allowing communication with other facilities and the output of information about themselves. The networking of all systems leads to 'cyber-physical production systems' and thus smart factories, in which production systems, components and people 'communicate via a network' and 'production is nearly autonomous'.

Schwab (2015) offers three reasons why 4.0 or 4IR (Schwab, 2018) represents a break rather than a prolongation of the third industrial revolution: velocity, scope and systems impact. It is unprecedented in terms of the speed of ongoing breakthroughs, evolving at an exponential rather than a linear pace, and it is disrupting almost every industry throughout the world, with the breadth and depth of these changes heralding 'the transformation of entire systems of production, management, and governance' (Schwab, 2015). He alludes to the unlimited 'possibilities of billions of people connected by mobile devices, with unprecedented processing power, storage capacity, and access to knowledge'.

In the Fourth Industrial Revolution, engineers, designers and architects combine computational design, additive manufacturing (joining materials to make objects from 3D model data), materials engineering (working with metals, ceramics, and plastics to create new materials) and synthetic biology (the application of engineering principles to biology to make biological components that do not already exist naturally). This is in order to 'pioneer a symbiosis between microorganisms, our bodies, the products we consume, and . . . the buildings we inhabit' (Schwab, 2015).

Writing for the Customer Relationship Management (CRM) platform, *Salesforce*, Devon McGinnis (2018) notes that 4IR comprises advances in artificial intelligence (AI), robotics and the Internet of Things (IoT). IoT, put simply, means connecting any device with an on and off switch to the Internet (and/or to each other, enabling them talk to each other). McGinnis gives some examples of the components of 4IR: 3D printing, genetic engineering and quantum computing. With respect to the last, according to a report in the *Financial Times*, researchers from Google have developed a computer that can perform a calculation in three minutes and 20 seconds, a feat that would take today's most advanced supercomputer, known as Summit, around 10,000 years (Giles, 2019).

The Fourth Industrial Revolution is, as McGinnis puts it, the collective force behind many products and services that are fast becoming indispensable to modern life:

> Think GPS systems that suggest the fastest route to a destination, voice-activated virtual assistants such as Apple's Siri, personalized Netflix recommendations, and Facebook's ability to recognize your face and tag you in a friend's photo.
>
> (McGinnis, 2018)

Public pedagogy exalting 4IR as a progressive capitalist force

Klaus Schwab

Like the preceding industrial revolutions, Schwab maintains, 4IR 'has the potential to raise global income levels and improve the quality of life for populations around the world'. 'To date', he stresses

> those who have gained the most from it have been consumers able to afford and access the digital world; technology has made possible new products and services that increase the efficiency and pleasure of our personal lives. Ordering a cab, booking a flight, buying a product, making a payment, listening to music, watching a film, or playing a game – any of these can now be done remotely.
>
> (Schwab, 2015)

Klaus Schwab happens to be the founder (in 1971) and Executive Chairman (sic) of the WEF that, according to Stephan Richter, publisher and editor-in-chief of *The Globalist* and Uwe Bott, its chief economist, 'has always naturally articulated the views of the world's economic elites' (Richter and Bott, 2019) or in the blunter words of Patrick Craven, writing in the *The Daily Maverick*, serves to 'bring together the world's rich and powerful capitalist leaders to discuss how best to protect their wealth and privileges' (Craven, 2017).

Thus it is not surprising that Schwab (2015) rests assured of the vast future benefits for capitalism that include technological innovation leading to 'a supply-side miracle, with long-term gains in efficiency and productivity'. In addition, transportation and communication costs will go down, while 'logistics and global supply chains will become more effective, and the cost of trade will diminish, all of which will open new markets and drive economic growth'.

Schwab (2015) does acknowledge the possibility of 'greater inequality' – 'the greatest societal concern associated with the Fourth Industrial Revolution' and 4IR's 'potential to disrupt labor markets' with 'the net displacement of workers by machines' possibly exacerbating 'the gap between returns to capital and returns to labor'. 'On the other hand', he notes reassuringly, 'it is also possible that the displacement of workers by technology will, in aggregate, result in a net increase in safe and rewarding jobs'. While the future may well be a combination of more inequality and more safety and rewards, Schwab (2015) is 'convinced of one thing – that in the future, talent, more than capital, will represent the critical factor of production'. At the opening of the 2013 WEF in Davos, Schwab went so far as to assert:

> Capital is being superseded by creativity and the ability to innovate – and therefore by human talents – as the most important factors of production. Just as capital replaced manual trades during the process of industrialization, capital is now giving way to human talent. Talentism is the new capitalism.
>
> (cited in Beach, 2014)

Talent, he maintains, will give rise to 'a job market increasingly segregated into "low-skill/low-pay" and "high-skill/high-pay" segments', leading to 'an increase in social tensions'. However, this can be mitigated. Thus, Schwab (2015) concludes with this clarion call:

> Neither technology nor the disruption that comes with it is an exogenous force over which humans have no control. All of us are responsible for guiding its evolution, in the decisions we make on a daily basis as citizens, consumers, and investors. We should thus grasp the opportunity and power we have to shape the Fourth Industrial Revolution and direct it toward a future that reflects our common objectives and values. . . . In the end, it all comes down to people and values. We need to shape a future that works for all of us by putting people first and empowering them.

Schwab's nod to nationalist populism

In a critique of Schwab and the economic elite that he represents, Stephan Richter and Uwe Bott (2019) consider how Schwab and 'the broader Davos set' are trying to adjust to the age of populism, specifically how the 'corporate elites are trying to accommodate nationalist populism'. They refer to an article Schwab (2018) wrote in the run-up to the 2019 annual meeting of the WEF in Davos, Switzerland. In the piece, Schwab began by acknowledging the fact that 'a substantial part of society has become disaffected and embittered, not only with politics and politicians, but also with globalization and the entire political system it underpins', and that in 'an era of widespread insecurity and frustration, populism has become increasingly attractive as an alternative to the status quo'. The 2019 WEF was one that Trump did not attend because of the partial shutdown of the US Government that occurred when he refused to approve a budget unless it included funds for a wall on the Mexican border. At the forum Schwab expressed disappointment in that Trump 'would have been and interesting discussion partner' (VOA, 2019).

In his 2018 article, Schwab had made a distinction between 'globalization' and 'globalism'. The former, he defined as 'a phenomenon driven by technology and the movement of ideas, people, and goods'; the latter as 'an ideology that prioritizes the neoliberal global order over national interests'. 'Nobody', he concluded, 'can deny that we are living in a globalized world', but 'whether all our policies should be "globalist" is highly debatable.'[1]

With Trump and his base in mind, Richter and Bott (2019) describe this 'attack on globalism' as 'a hijacking attempt'. 'Globalism', they point out, has been portrayed by right-wing populists as 'the root of all evil'. As I argued in Cole, 2019c, 69), following Ben Shapiro (2016), anti-globalists believe in conspiracy theories, particularly that those who oppose Trump's agenda must disagree with his premise that America's interests come first, and are therefore part of some sort of global conspiracy to overthrow American power. As Tim Squirrell (2017) explains:

> Their hyperbolic conspiratorial language might sound absurd, but it's become an increasingly coherent and important part of The_Donald' [a pro-Trump subreddit on Reddit]. Among their most common words are 'globalist scum'; 'the establishment', 'puppets', 'elites', 'masters' and 'cultural Marxist'.[2]

By advocating globalization over globalism, Schwab is able to both portray the former as 'ideology-neutral' while still maximizing the personal gains of the global economic elite (Richter and Bott, 2019). In their own words:

> In separating international exchange from global governance – and only attacking the latter . . . [Schwab] is encouraging a world in which politics is reduced to playacting action while economic elites remain unconstrained in their pursuit of profit. Schwab's rhetoric might make his Davos confab more palatable to Trump-style populists. But it does not do anything to solve the real problems that fuel political discontent.
>
> (Richter and Bott, 2019)

Schwab makes the case that it is 'not a matter of free trade or protectionism, technology or jobs, immigration or protecting citizens, and growth or equality', and declares these 'false dichotomies', insisting that we can avoid them by 'developing policies that favour "and" over "or"' and 'allowing all sets of interests to be pursued in parallel' (Schwab, 2018). In so doing, Schwab can easily cosy up to Trump. Europe editor for *Quartz* Lianna Brinded (2018) reveals how Schwab's appeasement manifests itself on a person-to-person basis. Side-by-side with the US President at the 2018 World Economic Forum, Schwab had introduced Trump with some congratulatory remarks that made his obsequious support clear:

> Mr. President, the role of the United States and your personal leadership is absolutely essential. For this reason, your message here has tremendous relevance. Of course, I'm aware that your strong leadership is open to misconceptions and biased interpretations, so it is so essential for us in the room to listen directly to you.
>
> (cited in Brinded, 2018)

Railing against both the cynicism of the economic elites and the economic ignorance of the populism of Trump and his ilk, Richter and Bott (2019) conclude with public pedagogy that epitomises liberal politics

> advancing the accountability of large corporations, our own democracies, and global institutions – is, in the best sense of the word, the progressive solution to the challenges we face at the national and international level. It stands in stark contrast to the regressive withdrawal from democratic frameworks domestically and at the international level favored by the populists. It is globalism at its best – and, instead of his shallow attempt to placate the forces of global populism, should deserve Schwab's praise.

Schwab responds to 'the "Greta Thunberg" effect'

I referred in the last chapter to the 'confrontation' between Thunberg and Trump at the 2020 World Economic Forum in January of that year (see pp. 15–16). In the month before, Schwab followed up his nod in the direction of Trump and nationalist populism at WEF 2019, with a response to what he referred to as 'the "Greta Thunberg" effect' (no doubt cognizant of what was likely to follow from Thunberg in Davos in January, 2020). This was a brazen attempt to capture the progressive politics of climate change activism for 'business leaders' who 'now have an incredible opportunity' to 'move beyond their legal obligations and uphold their duty to society' (Schwab, 2019). Schwab began the article, published in *Project Syndicate* a left-centre online publication promoting 'progressive capitalism', with an assumption and an assertion that there is no alternative to world capitalism: 'What kind of capitalism do we want? He stated that the question may be the defining one 'of our era'. He went on to refer to three possible models, 'shareholder capitalism', 'embraced by most Western corporations, which holds that a corporation's primary goal should be to maximize its profits'; '"state capitalism," which entrusts the government with setting the direction of the economy, and has risen to prominence in many emerging markets'; and 'stakeholder capitalism', that he said he proposed fifty years ago and that 'positions private corporations as trustees of society.' He believes the stakeholder option 'is clearly the best response to today's social and environmental challenges' (Schwab, 2019). This represents a further attempt to favour 'and' over 'or' allowing all sets of interests to be pursued together (Schwab, 2018). In this case, capitalists carry on making profits, but capitalism makes concessions to 'shared goals, such as those outlined in the Paris climate agreement and the United Nations Sustainable Development Agenda' (Schwab, 2019). There is an added bonus to leaving 'their mark on the world' because 'millennials and Generation Z no longer want to work for, invest in, or buy from companies that lack values beyond maximizing shareholder value' and 'long-term success is closely linked to that of their customers, employees, and suppliers . . . as more investors look for ways to link environmental and societal benefits to financial returns' (Schwab, 2019) and thus the best scenario for profit-making in current circumstances. Schwab has jumped on the progressive capitalist ideological bandwagon of the likes of Richter and Bott in advocating that companies 'should pay their fair share of taxes, show zero tolerance for corruption, and 'uphold human rights throughout their global supply chains' (Schwab, 2019). There is further mention of 'shared value creation' that now includes '"environmental, social, and governance" (ESG) goals as a complement to standard financial metrics' (Schwab, 2019). Finally, executive pay should not skyrocket in order to align management decision-making with shareholder interests but should align instead with 'long-term shared value creation'; and while 'all companies should . . . maintain an entrepreneurial mindset', they 'should also work with other stakeholders to improve the state of the world in which they are operating' which 'should be their ultimate purpose' (Schwab, 2019). Schwab ends with the rhetorical question, 'Is there any other way'? This is a blatant sleight of hand. Schwab, currently Honorary Professor at the Geneva School of Economics and Management, knows full well that the three models he has put on the table are not exhaustive, especially in the current climate of a renewal of interest in (eco-) socialism, as discussed in the last chapter of this book.

In the space of just a year, Schwab attempted to appease both Trump and his base, and Thunberg and her many supporters in the interests of the continuance of world capitalism. In making the case for stakeholder capitalism in preference to shareholder and state capitalism, he has deliberately refused to acknowledge the ecosocialist alternative. In the final chapter, I address this alternative, but first I undertake a Marxist critique of the Fourth Industrial Revolution.

2.2 The Fourth Industrial Revolution (4IR) – Marxist critique

Marxists argue that not only do we need to assess the negative effects of 4IR (or more accurately Capitalism 4.0 – the most apt nomenclature for 4IR *under global capitalism*) we must also explore how it will affect the balance between capital and labour (Hughes and Southern, 2019). From a Marxist perspective, there is a fundamental flaw in the arguments of both Schwab, on the one hand, and Richter and Bott, on the other. Contrary to the former's claims, we are not all in control of and responsible for 4IR, nor are we able, under capitalism, to grasp power and shape the Fourth Industrial Revolution, directing it towards a future that reflects our common objectives and values. Under capitalism, it is not possible that the future can be made to work for all of us, in Schwab's (2015) words, by 'putting people first and empowering them', even though 'stakeholder capitalism', if it were possible to achieve (a big 'if'), is potentially less socially and environmentally destructive than 'shareholder capitalism' that has more blatant and obvious corporate greed as its core.

While Richter and Bott (2019) are surely correct in asserting that Schwab's (2018) rhetoric, far from being ideologically neutral, serves to promote the interests of the economic elite, their 'progressive' solution, namely to challenge the nationalists like Trump and make large corporations, democracies and global institutions all 'accountable', while a laudable reformist 'corrective,' that was picked up in Schwab (2019), cannot provide an equitable future for all in the Fourth Industrial Revolution.

While at first sight, as Craven (2017) suggests, the Fourth Industrial Revolution could be seen as a huge advance for humankind, 'raising the possibility of thousands of dangerous, unhealthy or even just boring jobs being performed by computers and machines, and workers being freed to enjoy more leisure time and live life to the full', the reality is very different. To take the US as an example, as research by Carl Benedikt Frey and Michael A. Osborne (2013, 48), co-directors of the programme on Technology & Employment at Oxford University reveal, is that about 47 percent of jobs are at risk as a result of computerisation within the next 20 years.

Relating 4IR to the rise of populism and nationalism, entrepreneur, philanthropist and 2020 Democratic presidential candidate, Andrew Yang suggested to *Hill.TV* in March 2019 that it was the fourth industrial revolution that, in part, 'led directly to Donald Trump being elected in that we automated away 4 million manufacturing jobs in Michigan, Ohio, Pennsylvania, Wisconsin, Missouri and Iowa' (Manchester, 2019). Revealingly, he went on:

> My friends in Silicon Valley know full well we're about to do the same thing to millions of retail workers, call center workers, truck drivers, and on and on through the economy. We need to think much bigger about how to help Americans handle this transition.

In addition to stealing workers' jobs, Capitalism 4.0 will widen levels of inequality even more. Currently the world's richest 1 percent, those with more than $1 million, own 44 percent of the world's wealth, while adults with less than $10,000 in wealth make up 56.6 percent of the world's population but hold less than 2 percent of global wealth (Inequality. org, 2019). Unsurprisingly, a report by the Swiss bank UBS (Treanor, 2016) has warned that the richest stand to benefit most from Capitalism 4.0. Axel Weber, chairman (sic) of UBS stated that inequality will increase not just between developed and developing countries, but within society between the rich and the poor (cited in Treanor, 2016). However, Capitalism 4.0 will have a more negative impact notably in parts of Latin America and India than in developed countries such as Switzerland, Singapore and the UK, as artificial intelligence and robots reduce the competitive advantage of cheap labour. The report, published in January, 2016, outlines a polarisation in the labour force and 'greater income inequality imply[ing] larger gains for those at the top of the income, skills and wealth spectrums' (Treanor, 2016).

Marx, value theory and technology

Both corporate elitist and liberal accounts fail to recognise that there is a basic antagonism inherent in capitalism. As Tom Hickey (2006, 192) has put it, capitalism has an inbuilt tendency to generate conflict. As he eloquently remarks, 'The objective interests of the bourgeoisie and the proletariat are incompatible, and therefore generate not a tendency to permanent hostility and open warfare but a permanent tendency toward them'. The interests of capitalists and workers are diametrically opposed, he writes elsewhere, since a benefit to the former (profits) is a cost to the latter (Hickey, 2002, 168). This is because the worker gets only a fraction of the value she or he creates (how much is dependent on historical and socio-economic factors, not least the extent to which workers are able to successfully demand higher wages) and the rest is appropriated, or hived off, by the capitalist. The wage does not represent the *total* value the worker creates. We *appear* to be paid for every single second we work. However, underneath this appearance, this fetishism, the working day (like under serfdom) is split in two: into socially necessary labour (and the wage represents this); and surplus labour, labour that is not reflected in the wage. While the value of the raw materials and of the depreciating machinery is simply passed on to the commodity in production, labour power is a peculiar, indeed unique, commodity, in that it creates new value (see Marx, 1887, especially chapter 1):

> [the] sole source of the capitalists' profits is the value added by the exploitation of the workers' unpaid labour and the value that this adds to the commodity they produce or the services they deliver, along with a market of consumers who can afford to buy them.
>
> (Craven, 2017)

Pushing down labour costs, increasing surplus value and ultimately profits, is thus the driving force of capitalism. Technology is pivotal in this process. As Carl Hughes and Alan Southern (2019, 61)) explain, following Harry Braverman (1974), new technologies are crucial in that, under capitalism, they serve capital's interests. It is important to point out that technology is not a by-product of capitalism. Rather it is *required* by capital in its relentless search for more and more profit. Technology allows for ever-increasing control over labour that in turn makes greater surplus value creation more viable. For capitalists, control over the workforce is vital in containing workers' ongoing attempts to increase their share of the value they create – hence the important role that pro-capitalist governments in forging anti-trade union legislation, in order to thwart these attempts.

The tendency of the rate of profit to fall

Each phase of each of the industrial revolutions has witnessed mechanisation replacing labour power, and more recently, under Capitalism 4.0, reasoning and mental capacities (Hughes and Southern, 2019, 61–62). Capitalism dictates an ongoing 'project of expelling labour power from the capitalist labour process through technological innovation' (Rikowski and Ocampo Gonzalez, 2018, 113). The ever-increasing technological drive for productivity, in order to undercut rivals by making commodities in the technosphere (not just a growing collection of technological hardware, but a system that encompasses all of the technological objects manufactured by humans) more cheaply, means more machines ('dead labour' – since labour produces them) and less labour power (the source of profit). Thus, there is a tendency of the rate of profit (the ratio of profit to investment) to fall (TRPF), meaning that booms get shorter and slumps, longer and longer and deeper and deeper. As Samir Hinks (2012) explains, because profit can only come from human labour:

> as more and more capitalists invest in the new machinery [in Capitalism 4.0, in the technocapitalist complex] the average labour time required to produce each commodity [hypertechno-commodity in Capitalism 4.0] falls. This is what makes the rate of profit fall, as the ratio of surplus value to investment falls across the whole system.[3]

As Marx (1894) [1966] explains, '*thanks to the growing application of machinery* and fixed capital [physical assets used but not used up in production] in general (emphasis added), and given that the same number of workers in the same time, i.e. with less labour, convert an ever-increasing quantity of raw and auxiliary materials into products, the capitalist mode of production has an inbuilt tendency for the rate of profit to fall. It is worth quoting Marx at length:

> [The capitalist] mode of production produces a progressive relative decrease of the variable capital [wages] as compared to the constant capital [value of goods and materials], and consequently a continuously rising organic composition of the total capital. The immediate result of this is that the rate of surplus-value, at the same, or even a rising, degree of labour exploitation, is represented by a continually falling general rate of profit. . . . The progressive tendency of the general rate of profit to fall is, therefore, just *an expression peculiar to the capitalist mode of production* of the progressive development of the social productivity of labour. This does not mean to say that the rate of profit

may not fall temporarily for other reasons. But proceeding from the nature of the capitalist mode of production, it is thereby proved logical necessity that in its development the general average rate of surplus-value must express itself in a falling general rate of profit. Since the mass of the employed living labour is continually on the decline as compared to the mass of materialised labour set in motion by it, *i.e.*, to the productively consumed means of production, it follows that the portion of living labour, unpaid and congealed in surplus-value, must also be continually on the decrease compared to the amount of value represented by the invested total capital. Since the ratio of the mass of surplus-value to the value of the invested total capital forms the rate of profit, this rate must constantly fall.

It is important to stress that the TRPF is only a *tendency* rather than a law. The solution to the TRPF for the capitalist is to attack workers' conditions, for example, by increasing hours without increasing pay; giving workers less breaks; keeping them under greater surveillance and by laying off workers on contracts and replacing them with workers on zero hours contracts at very low rates of pay or the legal minimum wage if there is one; and upping output targets the more experienced the workers become. This allows the capitalist enterprise to both increase production and to sack workers who do not meet these targets to obviate the need to promote them to higher wages and/or increased benefits (examples are given later in this chapter in the context of Amazonization).

Under Capitalism 4.0, because of the 'rapid change of pace', the replacing of labour by computers and other machines increases and the contradictions multiply. As Craven (2017) explains, this acceleration and intensification means that the first companies who innovate with newer and more advanced machines make a big and quick profit by undercutting the prices of their competitors. But soon these competitors are forced to follow suit, and, because the surplus value added by fewer workers is smaller, the rate of profit tumbles. This was the case in the previous industrial revolutions, but following suit is sooner and profit crashes quicker in Capitalism 4.0. Moreover, as Craven continues: 'Meanwhile the thousands of workers who have been replaced by the machines do not have the money to buy the goods and services the machines are churning out'. He cites Nicolas Yan (2016) as quipping, 'After all, machines don't consume like humans do – a burger-flipping bot cannot enjoy a Big Mac, nor would a droid on the factory floor ever desire to purchase the iPhone that it assembles'. This leads to over-production of goods which further pushes down prices and profits.

However, it is imperative to point out that crises in capitalism do not derive from problems intrinsic to technology. As Hughes and Southern (2019, 62) insist:

If machinery was to be introduced in a truly collectivist society in which wealth was distributed fairly between citizens, it would have benefits both in reducing the amount of work that is necessary for production, while simultaneously increasing the wealth of both society and each individual member.

I make the case for a collectivist ecosocialist world in the last chapter of this book.

Capitalism 4.0 and gender

The first thing to stress is that the forms of knowledge that dominate Capitalism 4.0 – science and technology – have long been dominated by men (Ni Loideain and Adams, 2019). However with female names, voices and characters, artificially intelligent Virtual Personal Assistants (VPAs) such as Alexa (Amazon's virtual assistant), Cortana (Microsoft's), and Siri (Apple's) appear to be decisively gendered female (Ni Loideain and Adams, 2019). As Nóra Ni Loideaian and Rachel Adams point out, gendering as female is produced through mythical female names (Greek in the case of Alexa and Cortana and Nordic is Siri's case (2019) and specifically through a 'female' voice that users 'find more comfortable to instruct and give orders to than a male voice' (2019). Furthermore, 'these communications are delivered by witty and flirtatious characters revealed through programmed responses to even the most perverse questions'. This gendering amounts to what they refer to as 'an *unbodied* assemblage of a normative female'; in Amazon's words 'always ready, connected and fast' (cited in Ni Loideain and Adams, 2019) 'to obey the command of her user, and with no recourse to refuse or say no'. As Katherine Cross (2016) puts it, VPAs offer 'perfect subservience and total availability . . . free of messy things like autonomy, emotion, and dignity' (cited in Ni Loideain and Adams, 2019).

In addition, many of the opportunities the Fourth Industrial Revolution is thought to offer are, as has been demonstrated in this chapter, internet based. Yet, as a recent study (Majama, 2019) showed, in the world's second biggest continent, Africa (World Population Review, 2020b) women tend to have less access to internet based technologies than do men. This means that the impact on women's lives and work opportunities becomes a critical concern (Adams, 2019).

As Rachel Adams (2019) explains, with the massive increase in automation, 'routine intensive occupations' (Frey and Osborne, 2013) such as call centre and secretarial work are replaced by computers and robots are being prepared for care-worker jobs. Such jobs are generally occupied by women. In addition, because of the burden of care and domestic duties women carry on top of paid work (developed in the next chapter) women have significantly less time than men to undertake further education and training, meaning they will not easily be able to boost their digital skills.

More generally, there is the 'social harm caused by reproducing gender stereotypes in design choices that portray women as secondary to men, particularly given concerns around the lack of gender equality in STEM (science, technology, engineering and maths) fields' (Ni Loideain and Adams, 2019).

The Amazonization of everything

Fundamental to Capitalism 4.0 is the 'Amazonization' of everything (Golumbia, 2015). This also facilitates high-tech surveillance by the state by companies such as Amazon (Schuppe, 2019). Amazonization intensifies exploitation, while high-tech surveillance which derives from Amazonization provides necessary increases in state control over the populace. I will conclude this chapter with a consideration of each.

Amazonization serves as an example *par excellence* of the relationship between value theory à la Marx and technology. Amazon's compelling and unyielding drive to replace labour with the smart machines produced under Capitalism 4.0 and the consequent tendency of the rate of profit to fall requires the exploitation of labour on a hitherto unknown scale

Amazon and Walmart

Dmitri Koteshov, software development and current technology analyst for *Medici*, a Fin Tech (financial technology) enabling organisation, defines 'Amazonization' as the large-scale disruption that occurred in the retail trade as a result of Amazon's game-changing method of operation (Koteshov, 2020). He draws attention to the fact that the term is not unique and follows the pattern of 'Walmartization'[4] that took North America by storm in the 1990s and early 2000s. 'Born in the first dot-com era, the growth of Amazon', he reveals, 'has been particularly exponential since 2010 as the retail giant has more than quintupled in value' (Koteshov, 2020).

Writing for *Jacobin*, David Golumbia (2015) pinpoints the way in which Amazon's software strategy uses, and markets or helps its partners to sell through its Web Services, 'enterprise-wide tools' similar to those used by Walmart to track and manage every aspect of the company's business. This ranges from where the products are located in the warehouse, to the time it takes warehouse staff to take them out of their bins, to the number of seconds spent by (potential) customers on a product's webpage, as well as the number of products left unpurchased when they log out. Simon Head, author of *Mindless: Why Smarter Machines Are Making Dumber Humans* (Head,

2014, cited in Golumbia, 2015) observes how Amazon equals Walmart in its use of monitoring techniques to track the minute-by-minute movements and performances of employees, based on scientific management (Taylor, 1911) or Taylorism (named after Frederick Winslow Taylor), of which Amazon's shop-floor processes are an extreme version.

Micromanaging, exploiting and diminishing its workforce

Amazon has considerable expertise in micromanaging the lives of its employees and contractors. 'If an employee is behind schedule', according to Head, 'she will receive a text message pointing this out and telling her to reach her targets or suffer the consequences' (cited in Golumbia, 2015). The longer employees work in the warehouses, the higher their 'output targets' go, and Head suggests that these targets allow Amazon to fire employees as they gain the seniority that might entitle them to better pay and benefits (Golumbia, 2015).

Like Walmart, Amazon is a major purveyor of purposeful precarity, using all the legal means it can to prevent the classification of its workers as full-time employees entitled to benefits, and to block workers from unionizing. Also, like Walmart, 'it employs thousands of people in near-sweatshop conditions, carefully skirting the edges of labor laws, and playing geographic domains off each other to exploit its employees to the fullest extent possible'(Golumbia, 2015). But like nobody else, Amazon is skilful at breaking down problems into tiny pieces, parcelling them out to people all over the world forcing them to work for next to nothing, while pitting these low-wage workers in completion with each other.

Amazon's core asset is 'its mastery of distribution channels, for which it relies on a network of enormous warehouses located all over the world' (Golumbia, 2015). Conditions for workers in these warehouses, Golumbia (2015) points out, are much worse than those of its software developers and business managers, and as a result, 'the warehouses have repeatedly been the target of exposés, protests, and unionization attempts'.

The company is contributing to the casualization and minimisation of labour in a number of ways. First, Amazon has long been aggressive in its pursuit of alternative delivery strategies, perhaps most famously its proposed use of delivery drones (Golumbia, 2015). In June, 2019, it debuted a 'new, fully electric version of its package delivery drone' that it said would start shipping packages to consumers 'in a matter of months' (D'Onfro, 2019). The drones will be able to fly up to 15 miles and deliver packages under five pounds in weight (75 and 90 percent of Amazon deliveries) to customers in less than 30 minutes, while their hybrid design allows them to take off and land vertically, but also fly horizontally. The US Federal

Aviation Administration has given Amazon one year approval for research and testing (D'Onfro, 2019).

Second, the company also announced details of new warehouse robots can take packages from where human workers place them and autonomously sort them, zipping off to a specific area where they'll be sent off to a given destination, as well as ways for Alexa to predict what users want. For example, after buying movie tickets the device will automatically prompt the user to make a restaurant reservation through OpenTable (an online restaurant-reservation service company) and order an Uber taxi, an interaction that takes on 13 interactions instead of 40, according to Rohit Prasa, Vice President and Head Scientist, Alexa Artificial Intelligence (D'Onfro, 2019).

Amazon founder, chief executive officer and president Jeff Bezos is the richest person in the world, with an estimated net worth of $115 billion. He is one of only three billionaires having in excess of $100 billion (Andrew, 2020).

Is Alexa listening in?

Writing for Bloomberg, Matt Day et al. (2019) point out, while tens of millions of people use devices like Alexa, millions more are wary of having the devices in their homes for fear that someone might be listening in. The reality is that Amazon employs thousands of people around the world to do just that. The recordings are transcribed, annotated and fed back into the software to eliminate gaps in Alexa's understanding of human speech and help her better respond to commands, all in the interests of making Alexa more and more efficient in encouraging consumer expenditure. The team, from the US, Costa Rica, India and Romania, apparently sign 'nondisclosure agreements barring them from speaking publicly about the program' (Day et al., 2019). Sometimes, the workers pick up recordings that are upsetting and possibly criminal, and while Amazon says it has procedures in place for workers to follow when they hear something distressing, two Romania-based employees said that, after requesting guidance for such cases, they were told it wasn't Amazon's job to interfere. A spokesperson stated:

> We have strict technical and operational safeguards, and have a zero tolerance policy for the abuse of our system. Employees do not have direct access to information that can identify the person or account as part of this workflow.
>
> (cited in Day et al., 2019)

However, a screenshot seen by *Bloomberg* shows that although the recordings sent to the Alexa reviewers do not provide a user's full name and address, they are associated with an account number, as well as the user's first name and the device's serial number.

While it is possible to opt out, the company says their recordings still might be analyzed by hand over the regular course of the review process. According to Amazon's website, no audio is stored unless Alexa detects the wake word or is activated by pressing the button. However, sometimes the device appears to begin recording without any prompt, 'and the audio files start with a blaring television or unintelligible noise' (Day et al., 2019). Whether or not the activation is mistaken, the reviewers must transcribe it. (Day et al., 2019).

Transforming the public into mini-entrepreneurs

Amazon has also announced that it may 'incentivize people to deliver packages to their neighbors', proposing to pay people to do this. As Golumbia (2015) comments:

> Amazon is proposing to pay people for an activity that is typically thought of as an ordinary part of community life. Most people are happy to take care of packages for their neighbors once in a while, and don't feel the need to be paid for these things or even necessarily want to be paid for them. But it's not a stretch to imagine capitalizing on this willingness by transforming people into entrepreneurial mini-distribution hubs, and depriving us of one of the few remaining opportunities for serendipitous, selfless interaction.

Initiatives like Alexa and Amazon's neighbour-delivery program suggest that the 'workplace' is becoming indistinguishable from the rest of our lives. 'Again and again', Golumbia (2015) comments 'Amazon demonstrates that its goal is to apply computational analytics to every part of the social sphere, and to wring whatever profits it can from low-margin, no-margin, and as-yet un-commoditized parts of the social world'.

High-tech surveillance by the state facilitated by companies such as Amazon

Trebor Scholz (2015) writes that Amazon's business plan amounts to 'wage theft and total workplace surveillance' (cited in Golumbia, 2015). Growing inequality under Capitalism 4.0 discussed earlier in this chapter requires much higher degrees of surveillance by the state more generally. While Amazon is known mainly for its online consumer marketplace, it is becoming a potent resource for federal, state and local authorities, 'peddling an array of tools that harness the power of cloud-based computing,[5] artificial intelligence and video analysis' (Schuppe, 2019). Indeed, Amazon's work for the (US) government has become a significant component of the company's move beyond e-commerce and into tools that run the internet, with its government contracts, through the company's cloud-computing subsidiary, Amazon Web Services, having ballooned from $200 million in 2014 to $2 billion in 2019. As Daniel Ives, who researches Amazon at the financial advisory firm Wedbush Securities, put it this has allowed the company to build 'a fortress of cloud deals' that include police departments, federal law enforcement, national intelligence agencies and immigration authorities (cited in Schuppe, 2019). Sharon Bradford Franklin, policy director of New America's Open Technology Institute, a non-profit organization that advocates for digital rights explains:

> While providing secure cloud storage does not appear to pose privacy threats, providing a package of technologies that includes powerful surveillance tools like facial recognition and doorbell cameras, plus the capability to pool data into a massive database and run data analytics, does create very real privacy threats.
>
> (cited in Schuppe, 2019)

For example, she went on, a homeowner might share voluntarily footage from their doorbell camera with police looking for clues for a crime that occurred nearby. The police might subsequently run the video through facial recognition software that is imperfect and might incorrectly identify innocent people as potential suspects. Those wrong matches could then be shared with other agencies, like immigration enforcement, with far-reaching consequences (Schuppe, 2019).

Jon Schuppe (2019) gives the example of Neighbors app, run by a company called 'Ring' and bought by Amazon for an estimated $1 billion in early 2018, that promotes a popular line of doorbell cameras by encouraging members to share videos of suspicious activity. The platform includes a law enforcement portal that allows police departments to post alerts and request video from members who live near where a crime has been committed. The portal can tell police agencies if anyone in a particular area recorded a Ring video at the time of a crime.

Chris Gilliard is a professor at Macomb Community College near Detroit who studies how the mass collection of data and the use of artificial intelligence to analyze it can lead to discrimination against minorities and the poor. He stated:

> Violent crime is down across the board in America, but the cameras and the links to the Neighbors app have the ability to ramp up fear of

invaders or people who don't belong in particular neighborhoods — and that fear could bring harm to people. That equals less safety.

(cited in Schuppe, 2019)

Amazon is not, of course, the only company promoting facial recognition software. According to some estimates provided by Kevin Reed, writing for the World Socialist Web Site (WSWS), the market for facial recognition systems will reach $7 billion by 2024 and, more generally, AI technologies $126 billion by 2025 (Reed, 2020a). For example, Clearview AI has developed a breakthrough app reportedly being used by over 600 law enforcement agencies across the US (Hill, 2020). As Kashmir Hill (2020) explains in the *New York Times*:

> The system – whose backbone is a database of more than three billion images that Clearview claims to have scraped from Facebook, YouTube, Venmo and millions of other websites – goes far beyond anything ever constructed by the United States government or Silicon Valley giants.

Reed (2020a) argues that one of the more frightening revelations of the *New York Times* report is that Clearview is developing 'an interface between the facial recognition app and augmented reality glasses'. Apparently this means that such a technology could enable law enforcement officers 'to walk down the street or into a crowd or demonstration' and everyone encountered 'would be identified in their field of vision in the same way that Facebook tags friends in photos posted to its social media platform' (Reed, 2020a).

All this has to be viewed in the context of the intensification of the class struggle in the US and, indeed worldwide, as part of a more aggressive deployment of biometric tools to build up the state surveillance apparatus (Reed, 2020a).

Amazon workers fight back

Amazon employees are drawing together demands for climate change awareness: on the use of artificial intelligence by the police and military; the company's participation in illegal attacks on immigrants; and are calling for a change in the working conditions of the super-exploited warehouse workers. In late January 2020, more than 350 workers, defying an external communications directive, spoke out publicly against corporate policies and in defence of fellow employees who had previously criticised Amazon's contribution to climate change. They also cited the company's complicity with the US Department of Homeland Security Immigration and Customs

Enforcement (ICE), its collaboration with the fossil fuel industry and its general mistreatment of its workers. Amazon had threatened to sack people for violating a newly updated communications policy by talking publicly about Amazon and climate change (Reed, 2020b). The group encouraged other Amazon workers to sign on to *Medium*, a social journalism and online publishing platform, stating that they would not publish any of the quotes until the total number reached at least 100 people. The endorsed statements included, 'I disagree with AWS [Amazon Web Services] enabling Palantir and ICE to surveil and separate children from their parents at the border'; and 'Amazon's supply chain should not be built at the expense of warehouse workers who work at a pace that causes higher-than-industry-average injury rates. It's not humane to have people scared to go to the bathroom' (cited in Reed, 2020b). A Tweet read:

> Hundreds of us decided to stand up to our employer, Amazon. We are scared. But we decided we couldn't live with ourselves if we let a policy silence us in the face of an issue of such moral gravity like the climate crisis.
>
> (cited in Reed, 2020b)

In September 2019, nearly 5000 employees in 25 cities in 14 countries had gone on strike to protest the company's lack of action on climate change. This included 3,000 workers from the company's Seattle headquarters, part of the week-long Global Climate Strike that involved 4 million people worldwide. Amazon Employees for Climate Justice demanded more aggressive targets than Bezos' pledge to meet the goals of the Paris climate accord, arguing that the company should be carbon neutral by 2030, and that it should cease its web services contracts that help energy companies accelerate hydrocarbon extraction and that it must stop funding politicians and lobbyists who are climate deniers (Reed, 2020b).

As Kevin Reed explains Amazon's cloud computing centres have a significant carbon footprint, delivering an estimated one billion packages to consumers each year in the US alone. In addition, it was reported in the *New York Times* that Amazon contributed financially to the Competitive Enterprise Institute, a think-tank that disputes the dangers of climate change on behalf of the fossil fuel industry (Reed, 2020b).

As Reed (2020b) concludes, the latest confrontation at Amazon forms part of a growing wave of opposition among tech workers to the policies and practices of the giant tech monopolies that include Google (e.g. Ghaffary, 2019) and Microsoft (e.g. Lecher, 2019).

In this chapter, I have analysed how the four industrial revolutions have successively ramped up the exploitation of workers, and contrasted this by the way Capitalism 4.0 is portrayed by its capitalist protagonists, both unapologetic and liberal. I went on to critique 4.0 from a Marxist perspective, relating it to value theory and the tendency of the rate of profit to fall. I concluded the chapter with an analysis of the super-exploitation inherent in Amazonization and the fightback from its workers that includes criticism of Amazon's contribution to climate change. Marxists and other socialists must use pedagogy, institutional and public, to warn of the dire consequences of and threats arising from Capitalism 4.0, and at the same time, make the case for using the potential gains and benefits of the Fourth Industrial Revolution to give workers more leisure, to increase their standard of living, to contribute to the common good and crucially to create an abundance of all the commodities that people actually need. This is now manifestly self-evidently possible and will be increasingly so as 4IR progresses, but only in the context of a collectivist ecosocialist society. Asked whether he thinks capitalism can go green or new technology can save us, ecosocialist activist and editor of the journal *Climate and Capitalism*, Ian Ian Angus replies:

> The issue is not whether abstract capitalism, the imaginary capitalism of economics textbooks, could under ideal circumstances function without fossil fuels. The issue is whether really existing capitalism, the system that dominates the world today, can make the necessary changes in time to prevent large parts of the world from being made uninhabitable. . . . So there is little point in discussing what an ideal capitalism might do with ideal technologies. The real profit system is a giant obstacle to environmental progress, and there will be no permanent solution so long as it reigns.
>
> (Angus, 2019)

Having discussed the deleterious of global capitalism on the planet in Chapter 1, and having provided an explanation and critique of the Fourth Industrial Revolution in Chapter 2, in the last chapter of the book, I turn to ecosocialism as a solution. In order to do this, I first discuss socialism that has actually existed and then make the case for the urgent need now for an ecosocialism that is ecofeminist and antiracist.

Notes

1 In September 2019, Trump used his address to the UN general assembly to deliver a nationalist manifesto, denouncing "globalism" and illegal immigration and promoting patriotism as a cure for the world's ills (Borger, 2019).
2 This last term merits a little wider explanation than the other widely used words on The_Donald (a pro-Trump social news aggregation), since 'cultural Marxism' enjoys a wide currency among the alt-right. As Jason Wilson (2015) argues, it is a

flexible phrase 'that can be tailored to fit with the obsessions of a range of right-wing actors'. For the alt-right, 'cultural Marxism' is invoked for whatever they disapprove of, such as the various constituencies of perceived 'enemies'.

3 Technocapitalism derives from the work of Suarez-Villa (2009).

4 Walmartization or the 'Walmart Effect' refers to the economic impact felt by local businesses when a large company like Walmart opens a location in the area. It usually manifests itself, because of its massive buying power, by forcing smaller retail firms out of business and reducing wages for competitors' employees. The presence of a Walmart store can hurt the business of smaller companies and lower wages for local workers (Kenton, 2019).

5 In the simplest terms, cloud computing means storing and accessing data and programs over the Internet instead of the computer's hard drive (Griffith, 2016).

3 Saving the planet and harnessing technology for the good of all
The case for ecosocialism

Introduction

In this final chapter I begin by looking at socialism, focusing on its nineteenth and twentieth variants, taking the Paris Commune of 1871 as an instance of the former, and the Russian Revolution of 1917 as an example of the latter. In addition to the achievements of each, I address Stalinism with respect to the Soviet Union, as well as the Soviet Union and ecology. I go on to address socialism of the twenty-first century, arguing that ecosocialism is the only credible solution to the threat of impending ecological disaster, and that socialists should change the nomenclature 'socialism' to 'ecosocialism', instead. I draw on the work of Michael Löwy, Leigh Brownhill and Terisa Turner and others in promoting the necessity of a 'great transition initiative' towards ecosocialism, one that is also ecofeminist, and inclusive of diverse women in all continents. I conclude with a brief look at cultural change, before posing the question as to whether we have arrived at a Gramscian 'moment'. By challenging the public pedagogies of the pro-capitalist lobby and appropriating for ecosocialism, the vast and very real potential gains that are heralded by the Fourth Industrial Revolution, ecosocialists can confidently make the case that lasting equality for humankind is becoming increasingly possible and achievable.

Socialism: some nineteenth- and twentieth-century examples

Given the almost universal recognition that we need to save the planet from climate change extinction, but scepticism about socialism, often nurtured in pro-capitalist public pedagogy, it is necessary to raise the issue of the achievability, viability and desirability of socialist society *per se*. The fundamental difference between socialism and capitalism was articulated by the late socialist feminist revolutionary, Nora Castañeda, founder of the Women's Development Bank of Venezuela, who once declared simply and directly, 'We are creating an economy at the service of human beings instead of human beings at the service of the economy' (cited in Magdoff and Magdoff, 2005).[1]

One of a number of questions, asked of Marxists and other socialists, 'ok, show me where socialism works in practice'.[2] As Alpesh Maisuria (2018, 308) has argued, '[w]orking against [capitalist] mystification and promoting a belief in the feasibility of alternatives to the neoliberal class-based status quo is probably the greatest task for critical educators, human rights advocates and Marxists'. As he puts it:

> The *common sense* that has prevailed, since the fall of the Berlin Wall, oscillates between: (1) *there is no alternative* (TINA) to the status quo; . . . (2) the alternatives to neoliberal capitalism are not feasible because they are less desirable and the status quo is *as good as it gets*; [and] (3) the socialist alternative is not feasible because socialism is idealist and utopian, and not practically realistic.
>
> (Maisuria, 2018, 309)

In an attempt to debunk these ideological constructions, I will provide two snapshots of actually existing socialism, focusing on examples from the nineteenth and twentieth centuries respectively.[3] It should be pointed out, however, that socialism has recently witnessed a revival in the global north, most notably when Jeremy Corbyn became leader of the Labour Party in the UK in 2015 and Bernie Sanders' 2016 presidential campaign in the US. Both have self-described as 'democratic socialists' (e.g. Calamur, 2015; Golshan, 2019) but it needs to be stressed, that despite right-wing political and media attempts to portray Corbyn as a 'Marxist,' and Sanders and Alexandria Ocasio-Cortez, and the other 'members' of the so-called 'squad' (Ilhan Omar, Ayanna Pressley and Rashida Tlaib) as 'socialists', their policies may more accurately be described as 'left social democratic.'[4] Despite these major reservations, public pedagogy on a large scale has brought the concept of 'socialism' to the mainstream.

The Paris Commune of 1871

Writing in the online journal *In Defence of Marxism*, Greg Oxley (2001) describes the Paris Commune, a revolutionary socialist government that ruled the capital of France from 18 March to 28 May 1871 temporarily replacing the capitalist state with its own organs of government, as 'one of the greatest and most inspiring episodes in the history of the working class'. The Commune was formed in the wake of the defeat of France in the Franco-German war (1870–1871) and the collapse of the Second Empire of Napoleon III (1852–1870), after the National Guard, a forerunner of the soviets of workers and soldiers' deputies which arose in the course of the revolutions in Russia in 1905 and 1907 (Oxley, 2001; see the next section of this chapter) and government soldiers joined forces and seized control. The newly elected Commune replaced the leadership of the National Guard as the official government of revolutionary Paris (Oxley, 2001). Oxley explains some of its key features:

> Under the Commune, all privileges for state functionaries were abolished, rents were frozen, abandoned workshops were placed under the control of the workers, measures were taken to limit night-work, to guarantee subsistence to the poor and the sick. The Commune declared its aim as 'ending the anarchic and ruinous competition between workers for the profit of the capitalists', and the 'dissemination of socialist ideals'. The National Guard was open to all able-bodied men, and organised along strictly democratic lines. Standing armies 'separate and apart from the people' were declared illegal. The Church was separated from the state. Religion was declared 'a private matter'. Homes and public buildings were requisitioned for the homeless. Public education was opened to all, as were the theatres and places of culture and learning. Foreign workers were considered as brothers and sisters, as soldiers for the 'universal republic of international labour'. Meetings took place day and night, where thousands of ordinary men and women debated how various aspects of social life should be organised in the interests of the 'common good'.

The army and police were replaced by the people themselves, who were already armed and organised, with every neighbourhood defending and policing itself. Most government officials were elected, not appointed, and all could be replaced immediately by being voted out. The officials agreed that they should not be paid more than the workers' wages, so there was no encouragement for someone to become a government official for money or position (The Spark, 1977). As a result:

> the government was no longer a power above the workers. It could represent the changing views of the workers, as new people replaced the old. Many workers were elected to the Commune, and many of the important government bodies were headed by workers. The business of government was carried on by the working people themselves. The workers proved that they could run a government, that they could run society.

> (The Spark, 1977)

One result was a big increase in efficiency and government was much cheaper. With services organised in the most efficient way, there were no big salaries paid to officials and no drain of money into private hands (The Spark, 1977).

The women of the Commune organised themselves to attack the problems they faced, fighting to get rid of prostitution, not by passing laws which punished the women, but by trying to make it economically unnecessary for any woman to be forced into prostitution. They attacked the idea that some children were 'illegitimate' and fought to make all children accepted on an equal footing (The Spark, 1977).

While, as Oxley (2001) points out, the social and political character of the society that was gradually taking shape was unmistakably socialist, the lack of any historical precedent, the absence of clear, organised leadership, the lack of a clear programme, combined with the social and economic dislocation of a besieged city necessarily meant that the workers fumbled cautiously forward in dealing with the concrete requirements of organizing society in their own interests. Marx and Engels, Oxley points out, were particularly critical of the communards' failure to take control of the Bank of France. However, as Marx (1871) acknowledges [sexist language was, of course, the norm at the time]:

> When the Paris Commune took the management of the revolution in its own hands; when plain working men for the first time dared to infringe upon the governmental privilege of their 'natural superiors', and, under circumstances of unexampled difficulty, performed it at [very low] salaries . . . the old world writhed in convulsions of rage at the sight of the Red Flag, the symbol of the Republic of Labor, floating over the Hôtel de Ville. And yet, this was the first revolution in which the working class was openly acknowledged as the only class capable of social initiative, even by the great bulk of the Paris middle class – shopkeepers, tradesmen, merchants – the wealthy capitalist alone excepted.

Oxley (2001) concludes: above all the Commune lacked time, and the process in the direction of socialism was cut short by the return of the Versailles army and the ensuing terrible bloodbath that put an end to the Commune when:

> the Versailles army entered Paris on 21st May 1871. . . . The communards fought with tremendous courage, but were gradually pushed towards the east of the city and finally defeated on the 28th May. [The President of the French Republic, Adolphe] Thiers' forces conducted a terrible slaughter of anything up to 30,000 men, women and children,

with perhaps another 20,000 killed in the following weeks. Firing squads were at work well into the month of June, killing anyone suspected of having cooperated in any way with the Commune.

(Oxley, 2001)

The Russian Revolution of 1917

In the twentieth century, the obvious example of a social revolution that brought socialism on a solid and relatively long-lasting basis is the Russian Revolution of October 1917, which overthrew the Russian monarchy and empire, removed Russia from the final phase of the First World War and created the Union of Soviet Socialist Republics (USSR), or Soviet Union. Writing for the World Socialist Web Site (WSWS), Andrea Peters (2019) explains how, when the working class of Russia overthrew the combined forces of Tsarism and capitalism, the revolution immediately faced enormous difficulties: World War I had physically devastated Russia, which was mired in poverty and backwardness; the social democrats of Europe had betrayed the struggles of their own working classes and the young revolution found itself isolated. As such, it had over vast territories, to take on the counter-revolutionary forces of imperialism, which sought to destroy the victory of Russia's masses and prevent the revolution from extending across the globe. However, against all odds the revolution prevailed (Peters, 2019). The new government, led by Vladimir Lenin, solidified its power after three years of civil war, which ended in 1920.

Achievements

Russian revolutionary socialist Victor Serge (1930) [2009] describes some of the key features undertaken in the first workers' state:

- food and articles of prime necessity were distributed among the workers at nominal prices, as a precursor to the proposed total abolition of money;
- public utilities were made free;
- all public services were free;
- rent was abolished;
- theatre tickets were distributed free among the workers by the unions and factory committees;
- the post and (in some towns) the tram service were also free;
- free meals for children were introduced;
- the proportion of land in Russia cultivated by the peasants, who became its owners, leaped from 54 to 96 percent;

- whole industries were nationalized, including mining, transport, electricity, oil, rubber and sugar;
- such a thirst for knowledge sprang up all over the country that new schools, adult courses, universities and workers' faculties were formed everywhere;
- the legal recognition of free union between the sexes;
- the easing of divorce;
- the legalization of abortion;
- the complete emancipation of women, and the ending of the authority of heads of families and of religious sanctions;
- prostitution never disappeared completely, but the disappearance of the rich classes who were its clients reduced it to relative insignificance;
- the Communist Party and the Council of People's Commissars affirmed repeatedly that the liberty of religious believers would not be interfered with.

Stalinism

However, in what is perhaps the greatest tragedy in the history of socialism, many of the revolution's great achievements were to be overturned by what became known as Stalinism. Joseph Stalin ruled the USSR from 1929 to 1953, transforming the country from a peasant society into an industrial and military superpower. However, he ruled by terror, and millions of his own citizens died during his dictatorship. After Lenin died in 1924, Stalin outmanoeuvred his rivals for control of the party. Once in power, he attempted to implement 'socialism' in one country, collectivized farming and had potential enemies executed or sent to forced labour camps. Stalin aligned with the United States and Britain in the Second World War (1939–1945) but afterwards there emerged a constant confrontation between Stalinism in the Soviet Union along with a number of Eastern European countries allied to the USSR, on the one hand and US capitalism and its allies, including Britain, on the other, with each trying to establish hegemony with threats of the deployment of nuclear weapons. This became known as the Cold War and lasted until the downfall of the Soviet Union in 1991. Other Stalinist countries, including the German Democratic Republic (East Germany), Poland, Czechoslovakia, Hungary, Romania, Bulgaria and Albania all collapsed in the late 1980s and early 1990s.

It should be stressed that the rise of Stalin was neither foreordained nor a 'natural' outgrowth of the October Revolution. In Andrea Peters' (2019) words, 'the Great Russian chauvinist and bureaucrat [Stalin] secured power in ferocious conflict with the proletariat, peasantry and cadre of the revolutionary socialist movement' (see Rogovin, 2019 for a detailed analysis). In his final years Lenin had anticipated the dangers posed by this bureaucratic tendency and fought against them. In this he was joined by Leon Trotsky, his co-leader of the Russian revolution. When Lenin was incapacitated by strokes in 1923 and died in 1924, Trotsky continued this struggle along with other members of the Bolshevik Party in the Left Opposition (LO) that had been formed in 1923 (Peters, 2019). In 1939, Stalin approved a final plan for Trotsky's murder and on the evening of 20 August, 1940, Trotsky was assassinated with an ice pick by Ramón Mercader, a Spanish communist whose mother, a loyal Stalinist, had put him up to the task of killing Trotsky at a house in the suburbs of Mexico City, where he had been offered asylum. Trotsky died from his wounds just over 24 hours later in hospital (Borger and Tuckman, 2017). The struggle against Stalinism continued right up until the end. In 1989, for example, a letter sent to Soviet leader Mikhail Gorbachev by a rank-and-file party member set off alarm bells. The letter writer described the Communist Party as being made up of 'opportunists,' 'elites' and bourgeois 'born-againers.' It called for the working class to 'take matters into its own hands as the head of its own party' in order to lead a 'class war.' In the same year, massive miners' strikes erupted in the country (Rogovin, 2019, cited in Peters, 2019). 'Armed with the knowledge of their own history', Peters concludes, 'Rogovin understood that the Soviet working class could be an unstoppable force'. The 'monstrous domestic policies implemented under Stalin, the growth of a privileged bureaucracy, the attack on social equality, the suppression of inner-party democracy', all flowed, Trotsky insisted, from 'Stalin's rejection of world revolution and promotion of Russian nationalism' (Peters, 2019). As Peters argues, Rogovin (2019) demonstrates how falling living standards and a growing gap between masses of workers and the privileged bureaucrats allied to the Stalin regime were the result of the spread of piecework (wages paid for according to the amount produced), the speeding up of production, food rationing in the cities, growing disparity in wages and punitive restrictions on labour turnover. This was all part of 'collectivization and Stalin's frenzied drive for industrialization based on fantastical demands that the Soviet Union outstrip its own planned development targets' (Peters, 2019).

The Soviet Union and ecology

While the role of development was clearly not ecologically sound, it would be simplistic to dismiss the Soviet era as simply representing ecocide, of which the Chernobyl nuclear accident of 1986 is the most potent symbol.[5] In fact 'Soviet ecology presents us with an extraordinary set of historical ironies' (Bellamy Foster, 2015). As Marxist ecologist, John Bellamy Foster (2015) explains, on the one hand, in the 1930s and 1940s, the USSR violently purged many of its leading ecological thinkers and gravely damaged the environment in its quest for rapid industrial expansion. On the other, the Soviet Union revolutionised climatology and introduced pioneering forms of conservation, seeking to preserve and expand its forests (Bellamy Foster, 2015).

According to Bellamy Foster, Soviet ecology can be divided into three periods: (1) early Soviet ecology, characterized by revolutionary ecological theories and key conservation initiatives, starting with the 1917 revolution, and continuing up to the mid-1930s; (2) the middle period, from late 1930s to the mid-1950s, dominated by Stalinist purges, rapid industrialization, the Second World War, the onset of the Cold War, but also aggressive reforestation; and (3) late Soviet ecology from the late 1950s to 1991, that witnessed the emergence of a powerful Soviet environmental movement – in response to the extreme environmental degradation in the decade following Stalin's death in 1953 (Bellamy Foster, 2015).

Ecosocialism in the twenty-first century

Ecosocialism to the mainstream

I referred earlier to socialism having been brought into the mainstream by the presence on the national stages of the 'left social democratic' politics of figures like Corbyn and Sanders and 'the squad'. Their influence has also helped to make 'eco-social democracy' more prominent on both sides of the Atlantic and beyond, complementing the grassroots movements for climate change awareness from below. After the intervention of Greta Thunberg in August 2018, followed by weeks and months of direct action by school climate strikers (FFF), and Extinction Rebellion sit-ins in the first quarter of 2019, Corbyn succeeded in getting an opposition day debate motion passed in parliament without a vote. It called for the declaration of climate change emergency and urgent remedial action such as a green industrial revolution as well as changes to transport, agriculture and other areas. He responded after the debate: 'We pledge to work as closely as possible with countries that are serious about ending the climate catastrophe and make clear to Donald Trump that he cannot ignore international agreements and action on the climate crisis' (cited in Walker, 2019).

Squad 'member' Ocasio-Cortez, for her part, released a vision of a Green New Deal that calls for 'net-zero' greenhouse gas emissions within 10 years through a massive green energy build-up, including high-speed rail and retrofitting every building, as well as universal health care and job guarantees. It is important to stress that her resolution specifically states that the Green New Deal bill should create 'appropriate ownership stakes' for the public, support 'community grants, public banks, and other public financing' In short, as Caroline Haskins (2019) points out, 'the resolution places public ownership front and center'. Nationalizing the energy industry under community management and control would clearly herald an important anticapitalist project (Löwy, 2018).

Whatever becomes of the proposed bill, and this, of course, relates directly to who becomes the next US president, Marxists have traditionally supported reforms that benefit the working class, as do ecosocialists, but with an overall vision of a sustainable world beyond capitalism. The immediate reform task is the phasing out of fossil fuels economy-wide, and their replacement with clean energy. This must entail finding alternative work for, and supporting affected workers from fossil fuel-dependent industries like petrochemicals, agribusiness and aviation, as well as much superfluous industrial production and, of course, the arms industry and the military (Dawson, 2019). Despite the end of the Cold War, nuclear annihilation remains on the agenda (see the Postscript to this book). As Dawson (2019) puts it, a 'campaign against militarism, and for a Global Green New Deal, would require challenging the intra-state economic and military competition upon which capitalism relies'. 'While the net effect', he goes on:

> will be to contract production, we must also invest in selective growth in certain sectors, from renewable energy to organic farming, as well as low-carbon, socially necessary activities such as education and the caring professions. This will involve reversing the neoliberal capitalist dogma that has imposed austerity for decades.
>
> (Dawson, 2019)

At the same time, we must continue to forge the connections between anticapitalism and socialism and ecology. Ian King, Sky News' business presenter, voices the commonplace pro-capitalist 'suspicion that many of these climate change protesters are merely anti-capitalist protesters sailing under a flag of convenience' (King, 2019). On the contrary, as Angus (2019) has argued,

> by calling ourselves ecosocialists we are saying that we don't view the environment as just one of many equally important concerns, just another stick to beat up capitalism with. Ecosocialists recognise the global environmental crisis as the most important problem that humanity faces in the 21st century.

'If socialists don't recognise its centrality', he insists, 'our politics will be irrelevant'. When, referring to the ongoing trajectory of capitalist development and the need for socialism, Marx (1852) famously proclaimed that people 'make their own history, but they do not make it as they please; they do not make it under self-selected circumstances, but under circumstances existing already, given and transmitted from the past', he did not have also impending environmental disaster in mind. However, that is now our reality: 'The way we build socialism, the kind of socialism we will be able to build, will be fundamentally shaped by the state of the planet we must build it on', and ecosocialism; 'in particular the Marxist wing of the ecosocialist movement . . . builds and acts on that understanding' (Angus, 2019).

Transitioning to ecosocialism

A socialism of the twenty-first century must not only reject the anti-democratic realities of the 'actually existing socialisms' of the twentieth, it must also challenge environmental obliviousness characteristic of most political thinking and policy-making in that century. The solution, in the words of Löwy (2018), is a 'great transition initiative' towards a transformative vision and praxis: ecosocialism.[6] Thus the core of ecosocialism is the concept of 'democratic ecological planning, wherein the population itself, not "the market" or a Politburo, make the main decisions about the economy'. Early in the Great Transition to a new way of life with its new mode of production and consumption, some sectors of the economy must be suppressed (e.g. the extraction of fossil fuels implicated in the climate crisis) or restructured. At the same time, new sectors need to be developed. Economic transformation must be accompanied by the active pursuit of full employment with equal conditions of work and wages. Such an egalitarian vision is essential both for building a just society and for engaging the support of the working class in the structural transformation of the productive forces (Löwy, 2018). Ultimately, he continues, this vision is incompatible with private control of the means of production and the planning process. In particular, for investments and technological innovation to serve the common good (congruent with the arguments put forward earlier in this book about the fruits of the Fourth Industrial Revolution channelled towards need and not profit), 'decision-making must be taken away from the banks and capitalist enterprises that currently dominate, and put in the public domain'. This will enable 'society itself, and neither a small oligarchy of property owners nor an elite of techno-bureaucrats', to 'democratically decide which productive lines are to be privileged, and how resources are to be invested in education, health, and culture' (Löwy, 2018). As Löwy argues, major decisions on investment priorities – such as terminating all coal-fired facilities or directing agricultural subsidies to organic production – should be taken by direct popular vote, while other, less important decisions could be taken by elected bodies, on the relevant national, regional, or local scale. Democratic decision-making on a local, national or even international level becomes easier and easier and more and more sophisticated as ongoing technological advances in the Fourth Industrial Revolution (4IR) bring people closer and closer together in various ways.

Löwy (2018) counters right-wing fearmongering that 'central planning' takes away freedom, since 'democratic ecological planning ultimately supports more freedom, not less'. He gives three reasons for this. First, it offers 'liberation from the reified "economic laws" of the capitalist system that shackle individuals'.

Löwy's second response to right-wing critics of central planning is that ecosocialism 'heralds a substantial increase in free time', since planning and the reduction of labour time 'are the two decisive steps towards what Marx (1894) called "the kingdom of freedom."' (Löwy, 2018). 'A significant increase of free time', must be 'a condition for the participation of working people in the democratic discussion and management of economy and of society' (Löwy, 2018). In this regard, Mary Mellor (e.g. 2018, 2019) has criticised Löwy and other ecosocialists for their neglect of gender issues. Specifically, Mellor (2019, 189) argues that many ecosocialists 'fail to recognize the role of reproductive work in mediating between nature and "the economy", through the daily regeneration of human (and non-human) life'. Even with the end of alienating paid work, she stresses, there would still be 'the unrelenting nature of care work throughout the life cycle' (Mellor, 2019, 189). Historically and for the most part, today, she correctly asserts, this labour has mainly been performed by women. The 'kingdom of freedom', hailed by Löwy (2018), she claims, is genderless (Mellor, 2019, 189). She then notes that 'socialist-feminists have long pointed out, in the dreams of a coming age in which one can hunt in the morning, fish in the afternoon and be a critic after dinner, there is never any mention of who cooks the dinner' (Mellor, 2019, 189).[7] Mellor proposes an ecofeminist model of 'sufficiency provisioning' (190). By 'sufficiency', she means an egalitarian organizing principle whereby 'sufficiency for one must be sufficiency for all' (190). 'Provisioning', she explains, is rooted in feminist economics and is concerned with both productive and reproductive labour. It is critical, she argues, in the development of 'a radical political economy that is both socially just and ecologically sustainable' (190). 'Provisioning', she maintains, is more comprehensive than standard political economics in that it encompasses a view of humans as 'bodily creatures' who are metabolically related to the environment and embedded in the natural conditions of the planet (191):

[Women] have a life cycle that requires nourishing, physically and emotionally. This requires constant work. Caring, cleaning, comforting, feeding, listening, watching, accompanying. There is no choice about this. If everyone is left to choose whether to carry out these activities people will die, get sick, starve, despair. Nor can this be left to democratic ecological planning. Embodied needs have to be met immediately. It is well recognized that by default this work falls mainly to women and girls whether as mothers, wives, daughters, or low-paid workers. If a red-green future will be bucolic freedom for the ex-workers/consumers, women's burden of work will never pass from them.

(Mellor, 2018)

Just as the capitalist mode of production 'treats natural resources and eco-systems – fossil fuels, water systems, forests, soils, the atmosphere, the climate system – as inexhaustible, "costless externalities"', so it also relies on a similar 'costless externality': the unpaid work of women in the gendered division of labour in reproducing the labour force with bodily needs and emotional sustenance outside the workplace (Mellor, 2019, 191). To which I would add, women also reproduce the next generation of female caregivers.[8] Moreover, under capitalism, women often perform a double shift (if they are also in paid labour), or a triple shift (for example, if they are also students).

A solution to this oppression in an ecosocialist future would be to expand the welfare state beyond such areas as nursing, teaching and caring 'into a public/social economy that ensures that everyone has the means of sustenance' (Mellor, 2018). Sufficiency provisioning thus implies two objectives: the public sector providing goods and services through 're-gendered and non-oppressive relations' in an expansion of high quality social provision and environmentally sustainability and social justice (Mellor, 2019, 192). Again developments in 4IR can facilitate this provision.

Löwy (2018)'s third counter to the right-wing accusation that 'central planning' diminishes freedom is that in fact democratic ecological planning represents a whole society's exercise of its freedom to control the decisions that affect its destiny. If capitalist democracy in theory prevents political decision-making power being confined to a small elite, why should this not also apply to economic decisions? (Kovel, 2002, 215; see also Bowles and Gintis, 1976).[9] Thus with ecosocialist central planning, production would be for need not profit:

Under capitalism, use-value – the worth of a product or service to well-being – exists only in the service of exchange-value, or value on the market. Thus, many products in contemporary society are socially useless, or designed for rapid turnover ('planned obsolescence'). By contrast, in a planned ecosocialist economy, use-value would be the only criteria for the production of goods and services, with far-reaching economic, social, and ecological consequences.

(Löwy, 2018)

Central planning, Löwy (2018) suggests, would focus on large-scale, rather than small-scale economic decisions: the decision, for example, to transform a plant from producing cars to producing buses and trams would be taken by 'society as a whole', but the internal organization and functioning of the enterprise would be democratically managed by its workers. While, there has been a lot of discussion about the 'centralized' or 'decentralized' nature of planning, the most important element is 'democratic control at all levels – local, regional, national, continental, or international'. Thus, planetary ecological issues such as global warming must be dealt with democratically on a global scale. 'This nested, democratic decision-making', Löwy (2018) argues, 'is quite the opposite of what is usually described, often dismissively, as "central planning," since decisions are not taken by any "center," but democratically decided by the affected population at the appropriate scale':

> Democratic and pluralist debate would occur at all levels. Through parties, platforms, or other political movements, varied propositions would be submitted to the people, and delegates would be elected accordingly. However, representative democracy must be complemented – and corrected – by Internet-enabled direct democracy, through which people choose – at the local, national, and, later, global level – among major social and ecological options. Should public transportation be free? Should the owners of private cars pay special taxes to subsidize public transportation? Should solar energy be subsidized in order to compete with fossil energy? Should the work week be reduced to 30 hours, 25 hours, or less, with the attendant reduction of production?
>
> (Löwy, 2018)

Stressing that the ecosocialist revolution is a process rather than an event, Löwy insists that the 'Great Transition' from capitalist destructive progress to ecosocialism 'is a historical process, a permanent revolutionary transformation of society, culture, and mindsets'. 'Enacting this transition', he goes on, 'leads not only to a new mode of production and an egalitarian and democratic society, but also to an alternative *mode of life*, a new ecosocialist *civilization*'. This goes 'beyond the reign of money, beyond consumption habits artificially produced by advertising, and beyond the unlimited production of commodities that are useless and/or harmful to the environment'.

'Such a transformative process' crucially 'depends on the active support of the vast majority of the population for an ecosocialist program'. 'The decisive factor', he concludes, 'in development of socialist consciousness and ecological awareness is the collective experience of struggle, from local and partial confrontations to the radical change of global society as a whole' (Löwy, 2018).

Transitioning to ecosocialism entails breaking through the 'bizarre ideological mechanism, [in which] *every* conceivable alternative to the market has been discredited by the collapse of Stalinism' (Callinicos, 2000, 122), whereby the fetishization of life makes capitalism seem natural and therefore unalterable and where the market mechanism 'has been hypostatized into a natural force unresponsive to human wishes' (Callinicos, 2000, 125). Capital presents itself 'determining the future as surely as the laws of nature make tides rise to lift boats (McMurtry, 2000, 2), 'as if it has now replaced the natural environment. It announces itself through its business leaders and politicians as coterminous with freedom, and indispensable to democracy such that any attack on capitalism as exploitative or hypocritical becomes an attack on world freedom and democracy itself' (McLaren, 2000, 32). As Callinicos puts it, despite the inevitable intense resistance from capital, the 'greatest obstacle to change is not . . . the revolt it would evoke from the privileged, but the belief that it is impossible' (2000, 128).

> Challenging this climate requires courage, imagination and willpower inspired by the injustice that surrounds us. Beneath the surface of our supposedly contented societies, these qualities are present in abundance. Once mobilized, they can turn the world upside down.
>
> (Callinicos, 2000, 129)

Ecosocialism must be ecofeminist

In 1997, Mellor described ecofeminism as 'a movement that sees a connection between the exploitation and degradation of the natural world and the subordination and oppression of women' (Mellor, 1997, 1). Specifically, it

> brings together the analysis of the ecological consequences of human 'progress' from the green movement, and the feminist critique of women's disproportionate responsibility for the costs and consequences of women's embodiment, to show how relations of inequality within the human community are reflected in destructive relations between humanity and the non-human world.
>
> (Mellor, 1997, viii)

Some 20 years later, Leigh Brownhill and Terisa Turner, members of the eco-feminist collective associated with the journal, *Capitalism Nature Socialism* insist, along with others writing in that journal (e.g. Barca, 2017; Feder, 2019; Giacomini et al., 2018, Kovel, 2005), that ecosocialism that is not ecofeminist is not worth its salt (Brownhill and Turner, 2020). Moreover, according to these writers, ecofeminism is inclusive of antiracism. Antiracism must, of course, be a key component of ecoscialism. In their words, ecofeminism is the recognition of and struggle against capitalists' racist colonization and exploitation of (that is, extraction of profits from) nature and women'. For Brownhill and Turner (2020, 1), in so far as ecofeminism 'is characterized by efforts to unite the exploited across historic social divisions (e.g. waged and unwaged)', it is '*the revolutionary way* to an ecosocialist, post-capitalist future'. It is the ecofeminist ecosocialism or ecosocialist ecofeminism (they use the terms interchangeably) of women-led indigenous peoples or of those women engaged in urban struggle, along with young people and people of colour, that 'reveal the depth of the crisis faced by humanity today and its resolution' rather than the 'discourse of gender equality within the bounds of neoliberal capitalism' (Brownhill and Turner, 2020, 3).

I agree unreservedly with Brownhill and Turner on the crucial role of the women they champion in the previous paragraph and their international significance and importance, especially in comparison with those who advocate equality under capitalism with no wish to transcend it (Brownhill and Turner, 2020, 3), and with Löwy (2020) that the 'workerist/industrialist dogmatism of the previous century is no longer current', I also concur with him as to the importance of trade unions (albeit significantly de-shackled from *merely* reformist objectives) and that, 'in the final analysis, we can't overcome the system without the active participation of workers in cities and countryside, who make up the majority of the population' (Löwy, 2020). As noted already in this book, and as stressed by Löwy (2020), ecological goals that relate to closing coal mines, oil wells, thermal power stations and so on *must* be accompanied by guaranteed employment for the workers involved.

Capitalism, ecosocialism and growth: ecofeminist ecosocialism in the global south

I began the second section of the first chapter of this book with a quote from Greta Thunberg, in which she talks about 'money and fairy tales of eternal economic growth'. As Löwy points out economic growth has not only divided capitalists and environmentalists, but also socialists and environmentalists. Eco-socialism, he maintains, rejects the dualistic frame of growth versus degrowth, development versus anti-development, because 'both positions share a purely *quantitative* conception of productive forces'. 'A third position', he argues, 'resonates more with the task ahead: the *qualitative transformation* of development' (Löwy, 2018):

> A new development paradigm means putting an end to the egregious waste of resources under capitalism, driven by large-scale production of useless and harmful products. The arms industry is, of course, a dramatic example, but, more generally, the primary purpose of many of the 'goods' produced – with their planned obsolescence – is to generate profit for large corporations. The issue is not excessive consumption in the abstract, but the prevalent *type* of consumption, based as it is on massive waste and the conspicuous and compulsive pursuit of novelties promoted by 'fashion.' A new society would orient production towards the satisfaction of authentic needs, including water, food, clothing, housing, and such basic services as health, education, transport, and culture.
>
> (Löwy, 2018)

The global south, whose basic needs are much greater than the global north, must of course pursue 'greater classical "development" – railroads, hospitals, sewage systems, and other infrastructure'. 'While many poorer countries will need to expand agricultural production to nourish hungry, growing populations', Löwy contends (following La Via Campesina, a worldwide network of peasant movements that has long argued for this type of agricultural transformation), 'the ecosocialist solution is to promote agroecology methods rooted in family units, cooperatives, or larger-scale collective farms – not the destructive industrialized agribusiness methods involving intensive inputs of pesticides, chemicals, and GMOs' (Löwy, 2018). La Via Campesina is fighting for:

- Food sovereignty
- Climate and environmental justice
- Peasants' rights
- Land, water and territories
- Dignity for migrants and waged workers
- Agroecology and peasants' seeds to facilitate international seed exchanges, and ensure an equitable sharing of the benefits arising from their use, and to respect farmers' rights so that they can continue to renew and pass on to future generations the full diversity of the millions of seeds they have selected and saved from generation to generation (La Via Campesina, 2019)
- International solidarity (La Via Campesina, 2020a)

On International Women's Day, 8 March 2020, La Via Campesina made the following proclamation in Harare. I will quote from it at length:

> La Via Campesina joins in solidarity and struggle with millions of peasants, workers, migrants, indigenous people, fishers, pastoralists, waged agricultural workers, housekeepers and people of all diversity among the rural and the urban working class, as well as other organized sectors, to denounce the capitalist and patriarchal system and its constant oppression of women and of our peoples. We condemn this perverse system that hides behind the malicious facade of conservative and authoritarian neoliberalism, exacerbating workers' precariousness, eroding our rights, eliminating public policy, and criminalizing and killing those who resist. Massive women's mobilizations on a global scale . . . as well as other actions and expressions of struggle confirm the urgency behind the demand to stop all forms of violence and oppression that women suffer throughout the world. . . . We also condemn 'patriarchal States' that strengthen and naturalize violence against women. . . . It is distressing to hear presidential declarations such as those of Jair Bolsonaro in Brazil or Lenin Moreno in Ecuador trivializing rape and harassment; denying facts, such as the elevated rates of underage pregnancies from rape and feminicides when the numbers worldwide are alarming and alert to a pandemic.[10] At the same time, they promote policies in favour of transnational companies creating more inequality, poverty and increasing the concentration of wealth in few hands. As peasant women . . . we also stand in rejection of the industrial farming model and its system of death that everyday poisons our land, and pose serious health risks to the women who work in the fields and to the people who consume such food.
>
> (La Via Campesina, 2020b)

La Via Campesina (2020b) went on to demand the enforcement of the UN Declaration on the Rights of Peasants. This guarantees the right to land, to fair prices, to production free of toxic agrochemicals and acknowledges the essential role that peasants play in the production of healthy food and in acting as guardians of common goods by promoting agroecology as the path towards Food Sovereignty and the cooling down of the planet. They concluded by reaffirming their commitment to the promotion of Popular Peasant Feminism:

> It is the path we have taken as a movement, born of the heat of the struggles for the defense of common goods, against imperialism, colonialism and patriarchy as the pillars that uphold a capitalist system of

death and destruction, through mining mega-projects and agribusiness, responsible of the violence and dispossession suffered by the women and peoples of the world.

(La Via Campesina, 2020b)

Brownhill and Turner (2019, 1, 4) advocate an ecofeminist politics of resistance in pursuit of 'ecofeminist ecosocialism' ('a global, horizontal, subsistence-oriented, decolonized communing political economy' (2019, 5). Referring to the continent of Africa, women there, they argue, 'joined, on an expanding scale, by diverse women of all continents', have been at the forefront of the struggle against global capitalism since the onset of neoliberalism in the 1980s. Today, they explain, these resistance politics have 'converged in the politics of transition to a fossil-fuel-free world' (Brownhill and Turner, 2019, 1):

> Being more fully and directly reliant on nature for their daily subsistence, specific African women have faced and resisted enclosure of their commons and collectively maintained indigenous knowledge, seeds, practices, food production, and energy technologies that offer clear alternatives to oil and petro-chemical reliant food and energy systems. The prominence of women in defending the commons against commodification has been evident in Africa for many decades.

In Kenya, by the late 1990s, activists were engaged in land occupations to transfer that land to the poor (Brownhill et al., 1997; Turner and Brownhill, 2001a, 2001b), while by the 2010s, a food sovereignty movement in East Africa was pursuing 'the same goals as did the farmers who ripped out coffee trees and planted indigenous bananas and vegetables thirty years before' (Brownhill and Turner, 2019, 1), linking food sovereignty to women's human rights and renewable solar energy (e.g. Brownhill et al., 2016).

In Nigeria, in the early 2000s peasant women carried on their earlier 1980s/1990s struggle, mobilizing to stop oil production and in 2002 occupied a Chevron oil export terminal, oil platforms, oil fields and oil flow stations in the context of global consumer boycotts of Exxon and Chevron Texaco seguing into huge protests throughout the world against George Bush's 'war for oil' in Iraq (Brownhill and Turner, 2019, 1, 3–4). By 1990, they point out, citing Nnimmo Bassey (2016), a new generation of environmental activists had taken up their mothers' struggles confronting Big Oil for 'system change, not climate change' through food sovereignty, agroecology and community based renewable energy (Brownhill and Turner, 2019, 4).

Thus both the anti-crop export struggles in East Africa anti-oil movements in Nigeria 'began as gendered struggles against capitalist resource-grabs' (Brownhill and Turner, 2019, 4). Ecosocialist transformation would, of course, end the exploitation of its resources by advanced industrial societies, as well as the massive debt system of the global south (Löwy, 2018).

Cultural change

The highly wasteful 'advertising industry' would have no place in ecosocialist transition and would be replaced by consumer associations that vet and disseminate information on goods and services, altering patterns of consumption being an ongoing educational challenge within a historical process of cultural change (Löwy, 2018). In the words of Ernest Mandel (1992, 206):

> The continual accumulation of more and more goods . . . is by no means a universal and even predominant feature of human behavior. The development of talents and inclinations for their own sake; the protection of health and life; care for children; the development of rich social relations . . . become major motivations once basic material needs have been satisfied.
>
> (cited in Löwy, 2018)

The transition to ecosocialism would necessarily 'confront tensions between the requirements of protecting the environment and meeting social needs, between ecological imperatives and the development of basic infrastructure, between popular consumer habits and the scarcity of resources, between communitarian and cosmopolitan impulses' (Löwy, 2018). Balancing such interests must become the task of a democratic planning process, one that has become liberated from the imperatives of capital and profit-making. As such the process can reach solutions 'through transparent, plural, and open public discourse', enabled by ongoing technological developments. Löwy (2018) concludes that such 'participatory democracy at all levels does not mean that there will not be mistakes, but it allows for the self-correction by the members of the social collectivity of its own mistakes'.

Moreover, as John Molyneux (2019) argues, socialist politics when seriously applied would strengthen, not weaken, both XR and the struggle against climate change. 'For a start', he goes on, 'it could help to identify those in the movements who are potential friends and allies' and 'its real opponents (the capitalist corporations and politicians who have a major vested interest in fossil fuels), even if they do appear with hypocritical declarations, unctuous smiles and patronising words of praise' (Molyneux, 2019).

A fundamental cultural change in an ecosocialist future must be a sharing of care work by *all* in the context of the expansion of the welfare state, as demanded by Mellor (2018) so that women are no longer expected to do all the caring in society, to foster in Simon Mair's (2018) words, a universal 'cultural ethic of care'. It should go without saying that ecosocialism of the twenty-first century should also be full inclusive of and promote full equality for all other identities as well as gender as outlined by Mellor and Brownhill and Turner earlier. In addition, a socialism that does not interrogate Stalinism will not make for healthy or convincing public pedagogy. Ecosocialist public pedagogy must also urgently respond to the ignorant and ahistorical caricaturing of scientific socialism (Engels, 1880) by the likes of Trump, Boris Johnson and their fellow travellers that currently go largely unchallenged.

A Gramscian moment?

Lest the foregoing account tempts us to allow optimism of the intellect and of the will to underestimate the resilience and staying power of capitalism, it should be stressed that, although, as we have seen, it was stalled for a number of years in the Soviet Union and in a number of other countries, capitalism survived, and indeed flourished in the first, second and third revolutions. On a positive note, however, that we may have arrived at a 'Gramscian moment' is reinforced by the crisis in neoliberalism (see Bramble, 2018; see also Cole, 2020c, 5–6) and the burgeoning global backlash against right-wing populism, fascism and Trump.

'Moment' is one of a number of theoretical concepts that can be attributed to Gramsci. According to Juan Dal Maso (2016), what Gramsci meant by 'moment' is 'a level of analysis that is part of a logical as well as historical progression, conceptual as well as political/strategic, open to diverse combinations and mediations'. Following Peter D. Thomas (2009, 241), Michel Filippini (2012, 7) argues that the Gramscian 'moment' is central for contemporary politics and that it contains at least two perspectives. First, there is a permanent perspective on the integral unity of the capitalist state form, its production of the 'political' in bourgeois society as a function of hegemonic relations, and the need to elaborate a proletarian hegemonic apparatus capable of challenging it with a power of 'a completely different type'. The Fourth Industrial Revolution and climate change are both historical and logical manifestations of the trajectory of capitalist society. Capitalism 4.0 may also, for the supporters of capitalism be seen as 'permanent,' in that, in Schwab's (2015) words, it represents a break rather than a prolongation of the third industrial revolution in its velocity, scope and systems impact, and the unprecedented way, in terms of the speed of ongoing innovations, that it is evolving at an exponential rather than a linear pace. Similarly, climate change is seen, almost universally, as unstoppable without unparalleled human intervention of 'a completely different type'. Both are open, as we have seen, to diverse combinations and mediations. For Marxists, the second perspective entails:

> A novel reformulation of Marxism as a 'philosophy of praxis', as a theoretical formulation of the perspectives of the united front and as the expansive philosophical form at last discovered with which to construct proletarian hegemony, 'renewing from head to toe the whole way of conceiving philosophy itself'.
>
> (Thomas, 2009, 241, cited in Filippini, 2012, 7)

In combining traditional socialism with climate change rebellion, ecosocialism also insists on 'system change' because of the need, as argued earlier by ecosocialist, Ashley Dawson (2019), for a democratically administered emergency program for ecological reconstruction, something that is completely irreconcilable with capitalism's private ownership of the means of production and the imperatives of profit maximization and growth. Ecosocialism inherently represents a counter-hegemonic force that is both conceptual, in that it adds a substantial and crucial component to 'pre-eco' socialism, and political/strategic since it facilitates alliances between left radical groups. In its recasting of Marxism in a head to toe way to encompass a critique of both the latest incarnation of capitalism – Capitalism 4.0 *and* climate change extinction – ecosocialism surely provides a novel reformulation of Marxist philosophy and a new basis for revolution.

The context of the possibility of a Gramscian moment is the world-wide escalations of the class struggle in a number of countries including France, Algeria, Germany, Belgium, Poland, Portugal, Israel, Iran, Egypt, Tunisia, South Africa, Sri Lanka, India, New Zealand, Sudan, Mexico, and the United States (Kishore, 2019).

As has been demonstrated throughout this book, the essential similarity between climate change and Capitalism 4.0 is velocity. Our challenge is to appropriate the Fourth Industrial Revolution for ecosocialism; to not only call a truce with Mother Nature, but to begin to work in harmony with the environment before it is too late.

The Left must strive to make clear not just the realities of the ongoing destructive effects of world capitalism on the environment, but also the necessary theoretical connections between capitalism, ecology, feminism and socialism, underlining the central role of women worldwide, both in the global south and north in doing this, as outlined earlier by Brownhill and Turner, Mellor and others. Struggles against capitalism and (impending) ecological catastrophe need to be game-changing in a clarion call for fundamental transition towards ecosocialism. As Ian Angus put it before the accelerating deterioration of recent years: 'There is a giant death sentence hanging over much of our world, and capitalism is the executioner . . . socialists must be ecosocialists, and humanity needs an ecosocialist revolution' (Angus, 2013).

Hans Schellnhuber, atmospheric physicist, climatologist and founding director of the Potsdam Institute for Climate Impact Research and former chair of the German Advisory Council on Global Change has said that Holocene (warming) climates are no longer accessible to Planet Earth, and we are moving towards conditions such as the mid-Pliocene, three million years ago (+2 °C – 3°C, 400–450 ppm), or the mid-Miocene, seventeen million years ago (+5 °C, as much as 500 ppm) (Schellnhuber, 2018). While we were out in the streets pressing for radical change in the 1970s, argues Kerryn Higgs, 'we still had time', now there is a slight chance that humans might adapt to mid-Pliocene, we have little hope of adapting to mid-Miocene (Higgs, 2018).

All this, of course, begs the question, 'Do we have any chance of winning the battle, before it is too late?' (Löwy, 2020). As Fred Magdoff, Emeritus Professor of Plant and Soil Science at the University of Vermont states bluntly, the doubters may well be right, but 'the alternative will not be pretty' (Magdoff, 2018). I will leave the last word to Löwy (2020):

> Unlike the so-called 'collapsologists' who clamorously proclaim that catastrophe is inevitable and that any resistance is futile, we think the future is open. There is no guarantee that this future will be ecosocialist: this is the object of a wager . . . in which we commit all our forces, in a 'labour for uncertainty'. But as Bertolt Brecht said, with grand and simple wisdom: 'Those who fight may lose. Those who don't fight have already lost'.

Notes

1 For an analysis of the many progressive developments in Venezuela, before and during the presidency of socialist Hugo Chávez, see Cole (2014).
2 For thirteen other common questions and objections concerning Marxism and socialism and a Marxist response, see Cole (2018, 278–288).
3 These examples draw from and develop Cole (2018, 288–293).
4 Jim Kavanagh (2020) has provided a description of Sanders with which I would concur, that is, as a

> moderate [I have said 'left', but it depends with whom you are comparing him and in what era] welfare-state Social Democrat, not a socialist or even anti-capitalist; anti-war with an historically anti-imperialist, but now

imperialist-accommodating, tinge; nominally independent but functionally an auxiliary Democrat; fiercely critical of Republicans but stubbornly shy about criticizing Democratic colleagues. He is also, I think, honest and trustworthy. You can see that he takes and fights for the positions he does because he believes in them, not because he is opportunistically pandering to a specific audience segment or to the donor class.

Kavanagh has a 'decidedly more leftist, socialist point of view', but fervently hopes Sanders will win the 2020 US General Election (very unlikely as of April 2020) and will vote for him if he gets the nomination, 'the first Democratic presidential candidate . . . [Kavanagh has] voted for in decades', since he represents 'a principled Social Democratic program to meet human needs, based on and supported by a mass movement', as opposed to 'a program of neoliberal tinkering to protect profit-making possibilities, based on and supported by capitalist donors/the donor class'(Kavanagh, 2020). For similar reasons, for many socialists in the UK, myself included, voting for Corbyn in the 2017 and 2019 UK General Elections was their first vote for the Labour Party for decades. With the defeat of Labour in 2019 and the victory of Boris Johnson, hopes of a progressive politics have been stalled. In the US there is still the hope that Trump may be defeated.

5 The use of the term 'ecocide' to refer to the Soviet Union was heavily coloured by widespread international use of the term, beginning in the 1970s as a criticism of the US deployment of the chemical 'agent orange' in its war on Vietnam (Bellamy Foster, 2015).

6 For a discussion of the intellectual roots of ecosocialism, see Löwy (2018).

7 Mellor (2019) is referring to famous quote from Marx in *The German Ideology:*

> as soon as the distribution of labour comes into being, each man has a particular, exclusive sphere of activity, which is forced upon him and from which he cannot escape. He is a hunter, a fisherman, a herdsman, or a critical critic, and must remain so if he does not want to lose his means of livelihood; while in communist society, where nobody has one exclusive sphere of activity but each can become accomplished in any branch he wishes, society regulates the general production and thus makes it possible for me to do one thing today and another tomorrow, to hunt in the morning, fish in the afternoon, rear cattle in the evening, criticise after dinner, just as I have a mind, without ever becoming hunter, fisherman, herdsman or critic.
>
> (Marx, 1845)

8 Mellor (2018) acknowledges that women's 'burden is also increasingly shared by men where longer lifespans and fewer children mean that many men become caregivers later in life'.

9 The inclusion of 'in theory' is this sentence is pivotal. Often in capitalist democracies there is little chance of fundamental changes occurring after elections. Maintaining the capitalist system remains a number one priority.

10 In January 2020, Moreno said at an investors' meeting that 'men are constantly subject to the danger of being accused of harassment'. He went on to state that it is good that women make harassment claims, but then claimed that 'sometimes they only target those people who are ugly for harassment' (cited in Charner and Alvarado, 2020). He asserted: 'If the person is attractive, based on the canons of society, [women] don't necessarily consider it harassment', receiving 'a couple of chuckles and applause from what appeared to be a mostly male

audience' (Charner and Alvarado, 2020). He later apologised, but the comments had already gone viral, prompting Quito's Council for the Protection of Rights, one of Ecuador's main human rights watchdogs, to comment:

> A sort of 'joke' about sexual harassment, which came from the country's highest authority, exposes how terribly naturalized this sexist act – which affects nearly every woman in different places and of any age group – is. Women are exposed to this type of violence regularly in educational centers, universities, workspaces, political organizations, public transportation, streets, plazas, etc. Maybe that's why it seems normal, tolerable and ridiculous.
>
> (cited in Charner and Alvarado, 2020)

Postscript
One hundred seconds to midnight

The time at the start of 2020 moved to one hundred seconds to midnight because the threats from climate change and nuclear war mean that humanity is closer to annihilation than ever before, according to the scientists behind the 'Doomsday Clock'[1] (Griffin, 2020). The clock was originally set in 1947 at seven minutes to midnight. The proximity was two minutes to midnight in 1953 during the Cold War (the first USSR Hydrogen bomb) and again in 2018 and 2019, the latter two also due to the twin threats of nuclear weapons and climate change. According to the Bulletin of Atomic Scientists, this record reflects increasing concern about the two existential threats, 'compounded by a threat multiplier, cyber-enabled information warfare' that undercuts society's ability to respond (cited in Griffin, 2020). 'The international security situation is dire', they expand, 'not just because these threats exist, but because world leaders have allowed the international political infrastructure for managing them to erode' (cited in Griffin, 2020). Ban Ki-moon, the former UN general-secretary who helped reveal the new time on the clock, wrote in *The Independent* that the change should serve as a 'wake-up call for the world': 'The decision to move the hands of the Doomsday Clock is backed by rigorous scientific scrutiny, and demands an equally rigorous multilateral response' (Ki-moon, 2020).

As the Bulletin puts it, underlining the analysis in Chapter 1 of this book, public awareness of the climate crisis grew over the course of 2019, mainly because of mass protests by young people around the world. At the same time, governmental action on climate change continues to fall far short of meeting the challenge at hand. At the 2019 UN climate meetings, 'national delegates made fine speeches but put forward few concrete plans to further limit the carbon dioxide emissions that are disrupting Earth's climate'. This limited political response happened during a year when the effects of human-caused climate change, as we also saw in Chapter 1, 'were manifested by one of the warmest years on record, extensive wildfires, and quicker-than-expected melting of glacial ice' (Bulletin of the Atomic Scientists, 2020).

'In the nuclear realm', they go on, 'national leaders have ended or undermined several major arms control treaties and negotiations during the last year, creating an environment conducive to a renewed nuclear arms race', and to 'the proliferation of nuclear weapons, and to lowered barriers to nuclear war'. At the same time political conflicts 'regarding nuclear programs in Iran and North Korea remain unresolved and are, if anything, worsening'. 'US-Russia cooperation on arms control and disarmament', they point out, 'is all but nonexistent' (Bulletin of the Atomic Scientists, 2020).

'Continued corruption of the information ecosphere on which democracy and public decision making depend', they warn, 'has heightened the nuclear and climate threats'. In 2019, 'many governments used cyber-enabled disinformation campaigns to sow distrust in institutions and among nations', thus 'undermining domestic and international efforts to foster peace and protect the planet' (Bulletin of the Atomic Scientists, 2020).

This is all, in turn, further compounded by the assertions of certain US politicians. 'Led by the United States', Damon informs us, 'the world's nuclear powers are massively expanding and modernizing their arsenals', as part of US preparations for what Defense Secretary Mark Esper called 'high-intensity conflicts against competitors such as Russia and China' (cited in Damon, 2020). As Damon concludes, 'the Trump White House is moving rapidly ahead with a $1 trillion plan to expand, "modernize" and miniaturize the US nuclear arsenal', with Elbridge A. Colby, one of the co-leads of the National Defense Strategy (2018) published by the Pentagon in January of 2018, commenting in an article in *Foreign Affairs*, entitled 'If You Want Peace, Prepare for Nuclear War' (Colby, 2018):

> The risks of nuclear brinkmanship may be enormous, but so is the payoff from gaining a nuclear advantage over an opponent. Any future confrontation with Russia or China could go nuclear. . . . In a harder-fought, more uncertain struggle, each combatant may be tempted to reach for the nuclear saber to up the ante and test the other side's resolve, or even just to keep fighting.
>
> (cited in Damon, 2020)

'The best way to avoid a nuclear war', Colby continued, 'is to be ready to fight a limited one.' In this dangerous world, he concluded, 'US officials' must demonstrate that 'the United States is prepared to conduct limited, effective nuclear operations' (cited in Damon, 2020).

Such dangerous and criminal public pedagogy is not confined only to the Trump administration. In January 2020, US Democratic Representative Adam Schiff, speaking during the second day of the impeachment trial of Trump, stated, 'the United States aids Ukraine and her people so that we can fight Russia over there and we don't have to fight Russia here' (cited in Damon, 2020). As Andre Damon points out, 'For most of the American population, the assertion that "we" are fighting Russia will come as a surprise', raising such a confrontation 'not just . . . a possibility, but . . . a statement of present fact'. If the US is already at war with Russia, then escalation in whatever form is rendered less of a quantum leap. Damon underlines the potential threat:

> The United States and Russia each possesses over 6,000 nuclear weapons. Just a fraction of these is sufficient to kill billions of people and destroy human society. A war between these two countries, in other words, would be a cataclysmic disaster.
>
> (Damon, 2020)

Ruling class bravado by the likes of Trump, Colby and Schiff is facilitated in large part by the nature of war in the Capitalism 4.0. David Barno and Nora Bensahel (2018) refer to 'a new generation of high tech weapons', informing us that 'the United States and some of its potential adversaries are incorporating the technologies of the Fourth Industrial Revolution into a range of innovative new weapons systems'. These include railguns (they do away with using conventional explosives to fire a projectile, instead using an incredibly powerful electric circuit that can be fired at 2.5 kilometres per second, over distances of up to 200 kilometres – Weedon, 2019); and directed energy weapons ('electromagnetic systems capable of converting chemical or electrical energy to radiated energy and focusing it on a target, resulting in physical damage that degrades, neutralizes, defeats, or destroys an adversarial capability' – Office of Naval Research, undated). There are also hyper-velocity projectiles (capable of turning 'the more than 40-year-old deck gun design into an effective and low-cost weapon against cruise missiles and larger . . . [pilotless] . . . aerial vehicles' – LaGrone, 2019) and hypersonic missiles. These missiles, according to R. Jeffery Smith (2019) are a revolutionary new type of weapon that has unprecedented ability to manoeuvre, and can strike almost any target in the world within a matter of minutes. Capable of 'traveling at more than 15 times the speed of sound, hypersonic missiles arrive at their targets in a . . . destructive flash, before any sonic booms or other meaningful warning' (Smith, 2019). So far, there are no sure-fire defences. 'Fast, effective, precise and unstoppable – these are rare but highly desired characteristics on the modern battlefield' and are being developed not only by the United States but by China, Russia and other countries. (Smith, 2019).

These new weapons, Barno and Bensahel (2018) conclude, 'will dramatically increase the speed, range, and destructive power of conventional weapons beyond anything previously imaginable'. To make matters even worse, on 1 February 2020, we learned that in January, the Pentagon had deployed a new smaller nuclear warhead aboard the ballistic missile submarine USS Tennessee as it sailed into the Atlantic in the midst of the spiralling crisis with Iran. Known as W76–2, it has an explosive yield of roughly five kilotons, a third of the destructive power of the 'Little Boy' bomb that killed some 140,000 people in Hiroshima in 1945. Each missile can be loaded with as many as eight such warheads, capable of hitting multiple targets (Van Auken, 2020).

A report by the Federation of American Scientists (FAS) strongly suggests that it is 'likely that the new low-yield weapon is intended to facilitate first-use of nuclear weapons against North Korea or Iran' (cited in Van Auken, 2020). As Bill Van Auken argues, the 'threat against Iran is part of far broader build-up to global war through which US imperialism is seeking to offset the erosion of its previously hegemonic domination of the global economy by resorting to the criminal use of overwhelming military force'. Trump's rhetorical threats to wipe out other countries have ceased to be mere rhetoric. Van Auken concludes:

> His threats to carry out the 'obliteration' of Iran and to rain 'fire and fury' upon North Korea are not merely hyperbole. The 'usable' nuclear weapons to commit such atrocities have already been placed in his hands.
>
> (Van Auken, 2020)

In the words of the Bulletin of the Atomic Scientists (2020), conflating the two existential threats to our world:

> Civilization-ending nuclear war – whether started by design, blunder, or simple miscommunication – is a genuine possibility. Climate change that could devastate the planet is undeniably happening. And for a variety of reasons that include a corrupted and manipulated media environment, democratic governments and other institutions that should be working to address these threats have failed to rise to the challenge. But given the inaction – and in too many cases counterproductive actions – of international leaders, the members of the Science and Security Board are compelled to declare a state of emergency that requires the immediate, focused, and unrelenting attention of the entire world. It is 100 seconds to midnight. The Clock continues to tick. Immediate action is required.

Note

1 The 'Doomsday Clock' has been maintained by the Bulletin of Atomic Scientists since 1947. Its intention is to symbolise the threat that humankind faces and how

close it is to doomsday, taking into account all of the scientific and technological threats to survival. The time on the 'clock' is announced every year (Griffin, 2020). Founded in 1945 by University of Chicago scientists who had helped to develop the first atomic weapons, the Bulletin uses the imagery of apocalypse (midnight) and the contemporary discourse of nuclear explosion (countdown to zero). The decision to move (or to leave in place) the minute hand is made in consultation with its Board of Sponsors, that includes 13 Nobel laureates. The 'clock' has become a universally recognized indicator of the world's vulnerability to catastrophe from nuclear weapons, climate change, and disruptive technologies in other domains (Mecklin, 2020 in Bulletin of the Atomic Scientists, 2020)

References

About #FridaysForFuture. 2019. Available at: www.fridaysforfuture.org/about

Abram, Nerilie J., McGregor, Helen V., Tierney, Jessica E., Evans, Michael N., McKay, Nicholas P., Kaufman, Darrell S. and the PAGES 2k Consortium. 2016. 'Early Onset of Industrial-era Warming Across the Oceans and Continents', *Nature* 536. Available at: www.nature.com/articles/nature19082

Adams, Rachel. 2019. 'The Fourth Industrial Revolution Risks Leaving Women Behind', *The Conversation*, 5 August. Available at: https://theconversation.com/the-fourth-industrial-revolution-risks-leaving-women-behind-121216

Agostinone-Wilson, Faith (ed.). 2020. *On the Question of Truth in the Era of Trump*, Boston, MA: Brill/Sense.

Ambrose, Jillian. 2020. 'Trump Weakened Environmental Laws After BP Lobbying', *The Guardian*, 23 January. Available at: www.theguardian.com/business/2020/jan/23/trump-weakened-environmental-laws-after-bp-lobbying?CMP=Share_iOSApp_Other

Andrew, Scottie. 2020. 'Jeff Bezos Is Still the Richest Person in the World', *CNN Business*, 1 January. Available at: https://amp.cnn.com/cnn/2020/01/01/business/jeff-bezos-2019-billionaire-index-trnd/index.html

Angus, Ian. 2013. 'Why We Need an Ecosocialist Revolution', *Climate and Capitalism*, 2 July. Available at: https://climateandcapitalism.com/2013/07/02/why-we-need-an-ecosocialist-revolution-2/

Angus, Ian. 2019. 'Ian Angus on the Politics of Ecosocialism', *Rebel*, 9 August. Available at: www.rebelnews.ie/2019/08/09/interview-with-ian-angus/

Association for Global New Thought (AGNT). Undated 1. Available at: www.agnt. org/about

Association for Global New Thought (AGNT). Undated 2. 'Our Common Future: Brundtland Report OVERVIEW'. Available at: www.agnrt.org/brundtland-summary

Australian Associated Press. 2019. 'Total Fire Bans in Six NSW Regions as 146 Blazes Burn Across the State', *The Guardian*, 1 December. Available at: www.theguardian.com/australia-news/2019/dec/01/total-fire-bans-in-six-nsw-regions-as-149-blazes-burn-across-the-state

Barca, Stefania. 2017. 'The Labor(s) of Degrowth', *Capitalism Nature Socialism* 30(2). Available at: www.tandfonline.com/doi/abs/10.1080/10455752.2017.1373300

Barno, David and Bensahel, Nora. 2018. 'War in the Fourth Industrial Revolution', *War on the Rocks*, 19 June. Available at: https://warontherocks.com/2018/06/war-in-the-fourth-industrial-revolution/

Bassey, Nnimmo. 2016. *Oil Politics: Echoes of Ecological Wars*. Montreal: Daraja Press.

Bawden, Tom. 2019. 'Greenland's Ice Sheet Is Melting Seven Times Faster than It Was 30 Years Ago', *The Independent*, 10 December. Available at: https://inews.co.uk/news/environment/greenland-ice-sheet-melting-latest-environment-climate-change-flooding-1337639

Baynes, Chris. 2019a. 'Greta Thunberg Says Trump's "Extreme" Climate Change Denial Is Helping Environmental Movement', *The Independent*, 13 November. Available at: www.independent.co.uk/environment/greta-thunberg-trump-climate-change-denial-environment-us-a9201521.html

Baynes, Chris. 2019b. 'Trump Administration Removes Quarter of All Climate Change References from Government Websites', *The Independent*, 25 July. Available at: www.independent.co.uk/news/world/americas/us-politics/trump-climate-change-government-websites-global-warming-a9020461.html

BBC News. 2015. 'Climate Change: Obama Unveils Clean Power Plan', 3 August. Available at: www.bbc.co.uk/news/world-us-canada-33753067

BBC News. 2019. 'Paris Climate Accords: US Notifies UN of Intention to Withdraw', 5 November. Available at: www.bbc.co.uk/news/world-us-canada-50297029

BBC News. 2020a. 'Immigration: No Visas for Low-skilled Workers, Government Says', 19 February. Available at: www.bbc.co.uk/news/uk-politics-51550421

BBC News. 2020b. 'Australia Fires: Death Toll Rises as Blazes Destroy 200 Homes', 1 January. Available at: www.bbc.co.uk/news/world-australia-50962728

BBC News. 2020c. 'Australia Fires: PM Admits Mistakes in Handling of Crisis', 12 January. Available at: www.bbc.co.uk/news/world-australia-51080567

Beach, Gary. 2014. '"Talentism" Is the New Capitalism', *The Wall Street Journal*, 17 July. Available at: https://blogs.wsj.com/cio/2014/07/17/talentism-is-the-new-capitalism/

Beamish, Claudia. 2020. 'We must raise our game in the run-up to COP26'. *Morning Star* January 14. Available at: https://morningstaronline.co.uk/article/f/we-must-our-game-run-cop26

Beck, Kevin. 2019. 'About the Four Types of Fossil Fuels', *Sciencing.com*, 6 September. Available at: https://sciencing.com/about-5403214-four-types-fossil-fuels.html

Bellamy Foster, John. 2015. 'Late Soviet Ecology and the Planetary Crisis', *Monthly Review*, 1 June. Available at: https://monthlyreview.org/2015/06/01/late-soviet-ecology-and-the-planetary-crisis/?v=79cba1185463

Binding, Lucia. 2020. 'Australia Bushfires: Smoke Drifts as Far as South America', *Sky News*, 7 January. Available at: https://news.sky.com/story/australia-bushfires-smoke-drifts-to-south-america-11903274

Borger, Julian. 2019. 'Donald Trump Denounces "Globalism" in Nationalist Address to UN', *The Guardian*, 24 September. Available at: www.theguardian.com/us-news/2019/sep/24/donald-trump-un-address-denounces-globalism

Borger, Julian and Tuckman, Jo. 2017. 'Bloodstained Ice Axe Used to Kill Trotsky Emerges After Decades in the Shadows', *The Guardian*, 13 September. Available at: www.theguardian.com/world/2017/sep/13/trotsky-ice-axe-murder-mexico-city

Bowles, Samuel and Gintis, Herbert. 1976. *Schooling in Capitalist America*, London: Routledge and Kegan Paul.

Bramble, Tom. 2018. 'The Crisis in Neoliberalism and Its Ramifications', *Marxist Left Review*, 16 Winter. Available at: http://marxistleftreview.org/index.php/no-16-summer-2018/162-the-crisis-in-neoliberalism-and-its-ramifications#_edn22

Braverman, Harry. 1974. *Labor and Monopoly Capital: The Degradation of Work in the Twentieth Century*, New York: Monthly Press.

Brinded, Lianna. 2018. 'Davos's Founder Defended Trump Against "Biased Interpretations"', *Quartz*, 26 January. Available at: https://qz.com/1190200/davos-2018-klaus-schwab-defends-donald-trump/

Brownhill, Leigh S., Wahu M. Kaara, and Terisa E. Turner. 1997. "Gender Relations and Sustainable Agriculture: Rural Women's Resistance to Structural Adjustment in Kenya." *Canadian Woman Studies/Les Cahiers de la Femme* 17 (2): 40–44.

Brownhill, Leigh and Turner, Terisa E. 2019. 'Ecofeminism at the Heart of Ecosocialism', *Capitalism Nature Socialism* 30(1). Available at: www.tandfonline.com/doi/full/10.1080/10455752.2019.1570650

Brownhill, Leigh and Turner, Terisa E.͘ 2020. 'Ecofeminist Ways, Ecosocialist Means: Life in the Post-capitalist Future', *Capitalism Nature Socialism* 31(1). Available at: www.tandfonline.com/doi/full/10.1080/10455752.2019.1710362?scroll=top&needAccess=true

Brownhill, Leigh, Wahu M. Kaara, and Terisa E. Turner. 2016. "Building Food Sovereignty through Ecofeminism in Kenya: From Export to Local Agricultural Value Chains." Canadian Woman Studies 31 (1/2): 106–112.

Bulletin of the Atomic Scientists. 2020. 'Closer than Ever: It Is 100 Seconds to Midnight', *Current Time*. Available at: https://thebulletin.org/doomsday-clock/current-time/

Bulman, May. 2019. 'Boris Johnson's "New Immigration Department" Plan Could Expand Hostile Environment, Experts Warn', *The Independent*, 15 December. Available at: www.independent.co.uk/news/uk/politics/boris-johnson-home-office-hostile-environment-department-immigration-borders-a9247951.html

Buncombe, Andrew. 2019. 'US Tells UN It Is Officially Pulling Out of Paris Agreement on Climate Change', *The Independent*, 4 November. Available at: www.independent.co.uk/news/world/americas/us-politics/paris-agreement-climate-change-us-un-accord-pact-a9185171.html

Calamur, Krishnadev. 2015. 'How Jeremy Corbyn Would Govern Britain', *The Atlantic*, 18 August. Available at: www.theatlantic.com/international/archive/2015/08/jeremy-corbyn-labour-britain/401492/

Callinicos, Alex. 2000. *Equality*, Oxford: Polity Press.

Carbon Tracker. 2013. 'Wasted Capital and Stranded Assets Press Release', 4 December. Available at: www.carbontracker.org/wasted-capital-and-stranded-assets-press-release/

Carrington, Damian. 2019. 'Climate Emergency: World "May Have Crossed Tipping Points"', *The Guardian*, 27 November. Available at: www.theguardian.com/

environment/2019/nov/27/climate-emergency-world-may-have-crossed-tipping-points?CMP=Share_iOSApp_Other

Carrington, Damian. 2020. 'Ocean Temperatures Hit Record High as Rate of Heating Accelerates', *The Guardian*, 13 January. Available at: www.theguardian.com/environment/2020/jan/13/ocean-temperatures-hit-record-high-as-rate-of-heating-accelerates?CMP=Share_iOSApp_Other

Cave, Damien. 2019. 'Australian Fire Officials Say the Worst Is Yet to Come', *The New York Times*, 11 November. Available at: www.nytimes.com/2019/11/11/world/australia/fires-sydney-new-south-wales.html

Charner, Flora and Alvarado, Abel. 2020. 'Ecuador's President Apologizes for Saying Women Only "Target" Ugly Men for Harassment Claims', *CNN*, 3 February. Available at: https://edition.cnn.com/2020/02/03/americas/ecuador-president-ugly-men-intl/index.html

Climate Emergency Declaration. 2020, 1 June. Available at: https://climateemergencydeclaration.org/climate-emergency-declarations-cover-15-million-citizens/

Cohen, Tim. 2020. 'Climate Crisis: A Teen and a President Face Off in Davos – But Around Them the World Is Changing', *Daily Maverick*, 22 January. Available at: www.dailymaverick.co.za/opinionista/2020-01-22-climate-crisis-a-teen-and-a-president-face-off-in-davos-but-around-them-the-world-is-changing/

Colby, Elbridge. 2018. 'If You Want Peace, Prepare for Nuclear War', *Foreign Affairs*, November/December. Available at: www.foreignaffairs.com/articles/china/2018-10-15/if-you-want-peace-prepare-nuclear-war

Cole, Mike. 2008. *Marxism and Educational Theory: Origins and Issues*, London and New York: Routledge.

Cole, Mike. 2014. 'The Bolivarian Republic of Venezuela: Education and Twenty-First Century Socialism', in Sara C. Motta and Mike Cole (eds.), *Constructing Twenty-First Century Socialism in Latin America: The Role of Radical Education*, New York: Palgrave Macmillan.

Cole, Mike. 2018. 'Social Class, Marxism and Socialism', in Mike Cole (ed.), *Education, Equality and Human Rights: Issues of Gender, 'Race', Sexuality, Disability and Social Class*, 4th edition, London and New York: Routledge.

Cole, Mike. 2019a. *Trump, the Alt-Right and Public Pedagogies of Hate and for Fascism: What Is to Be Done?* London and New York: Routledge.

Cole, Mike. 2019b. 'We Must Never Forget Theresa May's Full Frontal Assault on Families Like Mine', *The Huffington Post*. Available at: www.huffingtonpost.co.uk/entry/theresa-may-minimum-income-requirement_uk_5cfa1db5e4b06af8b506823e

Cole, Mike. 2020a. 'Afterword: Ecological Catastrophe in the Fourth Industrial Revolution: The Case for Eco-Socialism and the Role of the Media', in Faith Agostinone-Wilson (ed.), *On the Question of Truth in the Era of Trump*, Boston, MA: Brill/Sense.

Cole, Mike. 2020b. *Theresa May, the Hostile Environment and Public Pedagogies of Hate and Threat: The Case for a Future Without Borders*, London: Routledge.

Cole, Mike. 2020c. 'Racism and Fascism in the Era of Donald J. Trump and the Alt-Right: Critical Race Theory and Socialism as Oppositional Forces', in Vernon Lee and Evelyn Shepherd W. Farmer (eds.), *Critical Race Theory in the Academy*, Charlotte, NC: Information Age Publishing.

Collier, Ian. 2020. 'Australia Bushfires: New Zealand Skies Turned Bright Orange by Smoke', *Sky News*, 6 January. Available at: http://news.sky.com/story/australia-bushfires-death-toll-rises-as-temperature-drop-offers-respite-11901369

Craven, Patrick. 2017. 'The "Fourth Industrial Revolution" – Or Socialist Revolution?', *Daily Maverick*, 5 January. Available at: www.dailymaverick.co.za/opinionista/2017-01-05-the-fourth-industrial-revolution-or-socialist-revolution/amp/

Cremin, L. A. 1976. *Public Education*, New York: Basic Books.

Cross, Katherine. 2016. 'When Robots Are an Instrument of Male Desire', *Medium*, 27 April. Available at: https://medium.com/the-establishment/ when-robots-are-an-instrument-of-male-desire-ad1567575a3d

Curry, Colleen. 2017. 'Climate Change Is Creating More Child Brides, Report Finds', *Global Citizen*, 27 November. Available at: www.globalcitizen.org/en/content/climate-change-is-creating-more-child-brides-repor/

Dal Maso, Juan. 2016. 'Gramsci's Three Moments of Hegemony', *Left Voice*. Available at: www.leftvoice.org/gramsci-s-three-moments-of-hegemony

Damon, Andre. 2020. 'Bulletin of the Atomic Scientists Warns of "Civilization-ending Nuclear War"', *World Socialist Web Site (WSWS)*, 24 January. Available at: www.wsws.org/en/articles/2020/01/24/pers-j24.html

Dawson, Ashley. 2019. 'We Can't Beat Climate Change Under Capitalism. Socialism Is the Only Way', *In These Times*, 15 April. Available at: http://inthesetimes.com/article/21837/socialism-anti-capitalism-economic-reform

Day, Matt, Turner, Giles and Drozdiak, Natalia. 2019. 'Amazon Workers Are Listening to What You Tell Alexa', *Bloomberg*, 10 April. Available at: www.bloomberg.com/news/articles/2019-04-10/is-anyone-listening-to-you-on-alexa-a-global-team-reviews-audio

Desoutter Industrial. 2019. 'Industrial Revolution – From Industry 1.0 to Industry 4.0', *Saint-Herblain, Loire-Atlantique: Desoutter Industrial*. Available at: www.desoutter tools.com/industry-4-0/news/503/industrial-revolution-from-industry-1-0-to-industry-4-0

deutschland.de. 2018. 'Seven Facts About Karl Marx', 27 April. Available at: www.deutschland.de/en/topic/knowledge/200-years-of-karl-marx-seven-facts

de Vries, Daniel. 2019. 'A Million Species Threatened with Extinction, UN-backed Report Warns', *World Socialist Web Site (WSWS)*, 14 May. Available at: www.wsws.org/en/articles/2019/05/14/spec-m14.html

D'Onfro, Jillian. 2019. 'Amazon's New Delivery Drone Will Start Shipping Packages "In a Matter of Months"', *Forbes*, 5 June. Available at: www.forbes.com/sites/jilliandonfro/2019/06/05/amazon-new-delivery-drone-remars-warehouse-robots-alexa-prediction/#73b1d7b8145f

Dunne, Daisy. 2019. 'Northern Hemisphere's Extreme Heatwave in 2018 "Impossible" Without Climate Change', *Carbon Brief*, 12 June. Available at: www.carbonbrief.org/northern-hemispheres-extreme-heatwave-in-2018-impossible-without-climate-change

Ellabban, Omar, Abu-Rub, Haitham and Blaabjerg, Frede. 2014. 'Renewable Energy Resources: Current Status, Future Prospects and Their Enabling Technology', *Renewable and Sustainable Energy Reviews* 39, November. Available at: www.sciencedirect.com/science/article/abs/pii/S1364032114005656

Ellis, Katherine. 2019. 'How Social Media Is Driving Climate Change Conversation', *Newswhip*, 26 September. Available at: www.newswhip.com/2019/09/social-media-is-driving-the-climate-change-conversation/

Ellsmoor, James. 2019. 'Trump Administration Rebrands Fossil Fuels as "Molecules of U.S. Freedom"', *Forbes*, 30 May. Available at: www.forbes.com/sites/jamesellsmoor/2019/05/30/trump-administration-rebrands-carbon-dioxide-as-molecules-of-u-s-freedom/#b7328fb3a24d

Ellsworth, Elizabeth. 2005. *Places of Learning: Media, Architecture, Pedagogy*, New York: Routledge.

Engels, Frederick. 1872 [1978]. 'On Authority', in *Marx-Engels Reader*, New York: W. W. Norton and Co., second edition, translated: Robert C. Tucker; transcribed: by Mike Lepore. Available at: www.marxists.org/archive/marx/works/1872/10/authority.htm

Engels, Frederick. 1880 [1970]. 'Socialism: Utopian and Scientific', in *Marx/Engels Selected Works*, Volume 3, Moscow: Progress Publishers. Available at: www.marxists.org/archive/marx/works/1880/soc-utop/index.htm

Extinction Rebellion. 2019. 'Extinction Rebellion Has Landed Stateside'. Available at: https://extinctionrebellion.us/

Feder, Helena. 2019. '"Never Waste a Good Crisis": An Interview with Mary Mellor', *Capitalism Nature Socialism* 30(4). Available at: www.tandfonline.com/doi/abs/10.1080/10455752.2018.1499787?journalCode=rcns20

Filippini, Michele. 2012. 'Towards a New Gramscian Moment', *Critical Sociology* 39(4).

Foer, Jonathan Safran. 2019. *We Are the Weather: Saving the Planet Begins at Breakfast*, London: Hamish Hamilton.

Frey, Carl Benedikt and Osborne, Michael. 2013. *The Future of Employment: How Susceptible Are Jobs to Computerisation?* Oxford: Oxford Martin School: University of Oxford. Available at: www.oxfordmartin.ox.ac.uk/downloads/academic/future-of-employment.pdf

García, Manuel Jr. 2019. 'CO2 and Climate Change, Old and New', *Counterpunch*, 23 December. Available at:www.counterpunch.org/2019/12/23/co2-and-climate-change-old-and-new/

Gayle, Damien, Taylor, Diane and Gentleman, Amelia. 2020. 'Jamaica Deportations "Must Be Halted Until Windrush Report Published"', *The Guardian*, 7 February. Available at: https://amp.theguardian.com/uk-news/2020/feb/07/jamaica-deportations-must-be-halted-until-windrush-report-published

Gentleman, Amelia. 2019a. 'The Tories Spout Empty Promises on Righting the Wrongs of Windrush', *The Guardian*, 11 December. Available at: www.theguardian.com/uk-news/2019/dec/11/tories-immigration-windrush-hostile-environment-labour-lib-dems

Gentleman, Amelia. 2019b. 'Three Generations of Windrush Family Struggling to Prove They Are British', *The Guardian*, 18 December. Available at: www.theguardian.com/uk-news/2019/dec/18/three-generations-of-windrush-family-struggling-to-prove-they-are-british

Gentleman, Amelia. 2019c. 'The Tories Spout Empty Promises on Righting the Wrongs of Windrush', *The Guardian*, 11 December. Available at: www.theguardian.com/uk-news/2019/dec/11/tories-immigration-windrush-hostile-environment-labour-lib-dems

Gentleman, Amelia, O'Carroll, Lisa, Walker, Peter and Brooks, Libby. 2020. 'MPs Vote to Drop Child Refugee Protections from Brexit Bill', *The Guardian*, 9 January. Available at: www.theguardian.com/world/2020/jan/08/mps-vote-to-drop-child-refugee-protections-from-brexit-bill

Ghaffary, Shirin. 2019. 'Google Employees Protest the Company's "Attempt to Silence Workers"', *Recode*, 22 November. Available at: www.vox.com/recode/2019/11/22/20978537/google-workers-suspension-employee-activists-protest

Giacomini, Terran, Turner, Terisa, Isla, Ana and Brownhill, Leigh. 2018. 'Ecofeminism against Capitalism and for the Commons', *Capitalism Nature Socialism* 29(1) Available at: www.tandfonline.com/doi/full/10.1080/10455752.2018.1429221

Giles, Martin. 2019. 'Google Researchers Have Reportedly Achieved "Quantum Supremacy"', *MIT Technology Review*, 20 September. Available at: www.technologyreview.com/f/614416/google-researchers-have-reportedly-achieved-quantum-supremacy/

Giroux, Henry A. 1998. 'Public Pedagogy and Rodent Politics: Cultural Studies and the Challenge of Disney', *Arizona Journal of Hispanic Cultural Studies* 2.

Giroux, Henry A. 2000. 'Public Pedagogy as Cultural Politics: Stuart Hall and the "Crisis" of culture', *Cultural Studies* 14.

Giroux, Henry A. 2010. *Hearts of Darkness: Torturing Children in the War on Terror*, London: Paradigm Publishers.

Global Justice Now. 2020. 'International Women's Day: Six Stories of Women Who've Defended the Environment and Resisted Corporate Power', 4 March. Available at: www.globaljustice.org.uk/blog/2020/mar/4/international-womens-day-six-stories-women-whove-defended-environment-and-resisted

Golshan, Tara. 2019. 'Bernie Sanders's Definition of Democratic Socialism, Explained', *Vox*, 12 June. Available at: www.vox.com/policy-and-politics/2019/6/12/18661708/bernie-sanders-definition-democratic-socialism-explained

Golumbia, David. 2015. 'The Amazonization of Everything', *Jacobin*. Available at: www.jacobinmag.com/2015/08/amazon-google-facebook-privacy-bezos/

Gramsci, Antonio. 1971. *Selections from Prison Notebooks*, London: Lawrence and Wishart.

Greshko, Michael and National Geographic Staff. 2019. 'What Are Mass Extinctions, and What Causes Them?' *National Geographic*, 26 September. Available at: https://relay.nationalgeographic.com/proxy/distribution/public/amp/science/prehistoric-world/mass-extinction

Griffin, Andrew. 2020. 'Doomsday Clock: Humanity Closer to Annihilation than Ever Before, Scientists Say', *The Independent*, 23 January. Available at: www.independent.co.uk/news/science/doomsday-clock-2020-coronavirus-midnight-time-nuclear-war-ai-climate-a9298926.html?amp

Griffith, Eric. 2016. 'What Is Cloud Computing?', *PC Magazine*, 3 May. Available at: https://uk.pcmag.com/networking-communications-software/16824/what-is-cloud-computing

Harvey, David. 2009. 'Organizing for the Anti-Capitalist Transition', *Reading Marx's Capital with David Harvey*, 16 December. Available at: http://davidharvey.org/2009/12/organizing-for-the-anti-capitalist-transition/

Harvey, David. 2010 'The Enigma of Capital and the Crisis This Time', *Reading Marx's Capital with David Harvey*, 30 August. Available at: http://davidharvey.org/2010/08/the-enigma-of-capital-and-the-crisis-this-time/

Harvey, Fiona. 2019. 'UN Calls for Push to Cut Greenhouse Gas Levels to Avoid Climate Chaos', *The Guardian*, 26 November. Available at: www.theguardian.com/environment/2019/nov/26/united-nations-global-effort-cut-emissions-stop-climate-chaos-2030?CMP=Share_iOSApp_Other

Haskins, Caroline. 2019. 'Alexandria Ocasio-Cortez's Green New Deal Should Nationalize Utilities', *Vice*, 7 February. Available at: www.vice.com/en_us/article/d3m3dx/alexandria-ocasio-cortezs-green-new-deal-should-nationalize-utilities

Head, Mike. 2020. 'Australian PM Declares "New Normal" of Climate Disasters', *World Socialist Web Site (WSWS)*, 13 January. Available at: www.wsws.org/en/articles/2020/01/13/morr-j13.html

Head, Simon. 2014. *Mindless: Why Smarter Machines Are Making Dumber Humans*, New York: Basic Books.

Hickey, Tom. 2002 'Class and Class Analysis for the Twenty-first Century', in Mike Cole (ed.), *Education, Equality and Human Rights: Issues of Gender, 'Race', Sexuality, Special Needs and Social Class*, London and New York: RoutledgeFalmer.

Hickey, Tom. 2006. '"Multitude" or "Class": Consistencies of Resistance, Sources of Hope', in Mike Cole (ed.), *Education, Equality and Human Rights: Issues of Gender, 'Race', Sexuality, Disability and Social Class*, 2nd edition, London and New York: Routledge.

Higgs, Kerryn. 2018. 'Do Red and Green Mix? A Roundtable', *Great Transition Initiative*, December. Available at: https://greattransition.org/roundtable/ecosocialism-kerryn-higgs

Hill, Kashmir. 2020. 'The Secretive Company That Might End Privacy as We Know It', *The New York Times*, 18 January. Available at: www.nytimes.com/2020/01/18/technology/clearview-privacy-facial-recognition.html?action=click&auth=login-email&login=email&module=Top%20Stories&pgtype=Homepage

Hinks, Samir Karnik. 2012. 'What Is the Tendency of the Rate of Profit to Fall?', *Socialist Review*, July/August 371. Available at: http://socialistreview.org.uk/371/what-tendency-rate-profit-fall

Holloway, John. 2010. 'We Are the Crisis of Capital', *Red Pepper*, 16 June. Available at: www.redpepper.org.uk/we-are-the-crisis-of-capital/

Hughes, Carl and Southern, Alan. 2019. 'The World of Work and the Crisis of Capitalism: Marx and the Fourth Industrial Revolution', *Journal of Classical Sociology* 19(1).

Hunziker, Robert. 2019a. 'The Amazon at a Tipping Point', 24 December. Available at: www.counterpunch.org/2019/12/24/the-amazon-at-a-tipping-point/

Hunziker, Robert. 2019b. 'Biosphere Collapse?', *Counterpunch*, 19 December. Available at: www.counterpunch.org/2019/12/19/biosphere-collapse/

Inequality.org. 2019. 'Global Inequality'. Available at: https://inequality.org/facts/global-inequality/#global-wealth-inequality

The Intergovernmental Panel on Climate Change. 2018. 'Global Warming of 1.5 °C'. Available at: www.ipcc.ch/sr15/

The Intergovernmental Panel on Climate Change. 2019. *The Ocean and Cryosphere in a Changing Climate*. Available at: https://report.ipcc.ch/srocc/pdf/SROCC_ SPM_Approved.pdf

International Union for Conservation of Nature (IUCN). 2019a. 'Ocean Deoxygenation'. Available at: www.iucn.org/theme/marine-and-polar/our-work/climate-change-and-oceans/ocean-deoxygenation

International Union for Conservation of Nature (IUCN). 2019b. 'Marine Life, Fisheries Increasingly Threatened as the Ocean Loses Oxygen – IUCN Report', 7 December. Available at: www.iucn.org/news/marine-and-polar/201912/marine-life-fisheries-increasingly-threatened-ocean-loses-oxygen-iucn-report

Kavanagh, Jim. 2020. 'The Party's Over: Bernie's Last Dance With the Dems', *Counterpunch*, 31 January. Available at: www.counterpunch.org/2020/01/31/the-partys-over-bernies-last-dance-with-the-dems/

Kenton, Will. 2019. 'Walmart Effect', *Investopedia*, 3 October. Available at: www.investopedia.com/terms/w/walmart-effect.asp

Ki-moon, Ban. 2020. 'The Doomsday Clock shows we are closer to global catastrophe than ever before – this needs to be a wake-up call for the world'. *The Independent*. Available at: https://www.independent.co.uk/voices/doomsday-clock-announce ment-midnight-nuclear-weapons-climate-crisis-australia-a9298921.html

King, Ian. 2019. 'Extinction Rebellion Were Naïve on the Tube – And in the City Too', *Sky News*, 18 October. Available at: https://news.sky.com/story/sky-views-extinction-rebellion-should-stop-treating-big-business-as-the-enemy-11838095

Kishore, Joseph. 2019. 'The Future Lies in Socialism', *World Socialist Web Site (WSWS)*, 13 May. Available at: www.wsws.org/en/articles/2019/05/13/mdjk-m13.html

Kitagawa, Kaori. 2017. 'Situating Preparedness Education Within Public Pedagogy', *Pedagogy, Culture & Society* 25(1). Available at: www.tandfonline.com/doi/full/10.1080/14681366.2016.1200660

Koteshov, Dmitri. 2020. 'Amazonization: A Forest of Ideas for Banking and Finance', *Medici*, 3 January. Available at: https://gomedici.com/amazonization-forest-of-ideas-for-banking-finance

Kovel, Joel. 2002. *Enemy of Nature: The End of Capitalism or the End of the World?* New York: Zed Books.

Kovel, Joel. 2005. 'The Ecofeminist Ground of Ecosocialism', *Capitalism Nature Socialism* 16(2). Available at: www.tandfonline.com/doi/abs/10.1080/10455750 500108146

Lacy, S. (ed.). 1995. *Mapping the Terrain: New Genre Public Art*, Seattle, WA: Bay Press.

LaGrone, Sam. 2019. 'Navy Quietly Fires 20 Hyper Velocity Projectiles Through Destroyer's Deckgun', *U.S. Naval Institute*, 8 January. Available at: https://news.usni.org/2019/01/08/navy-quietly-fires-20-hyper-velocity-projectiles-destroyers-deckgun

La Via Campesina. 2019. 'Don't Give Up the International Seed Treaty to the New Genetic Biopiracy!', 19 November. Available at: https://viacampesina.org/en/dont-give-up-the-international-seed-treaty-to-the-new-genetic-biopiracy/

La Via Campesina. 2020a. 'What We Are Fighting For'. Available at: https://via campesina.org/en/what-are-we-fighting-for/

La Via Campesina. 2020b. '#8March2020: Peasant and Rural Women, Organised for Food Sovereignty and a Dignified Life!', 21 February. Available at: https://viacampesina.org/en/8march2020-peasant-and-rural-women-organised-for-food-sovereignty-and-a-dignified-life/

Lecher, Colin. 2019. 'Microsoft Employees Are Protesting the Company's "Complicity in the Climate Crisis"', *The Verge*, 19 September. Available at: www.theverge.com/2019/9/19/20874081/microsoft-employees-climate-change-letter-protest

Lenton, Timothy M., Rockström, Johan, Gaffney, Owen, Rahmstorf, Stefan, Richardson, Katherine, Steffan, Will and Schellnhuber, Hans Joachim. 2019. 'Climate Tipping Points – Too Risky to Bet Against', *Nature*, 27 November. Available at: www.nature.com/articles/d41586-019-03595-0

Lewontin, R. and Levins, R. 1997. 'Organism and Environment', *Capitalism Nature Socialism* 8(2).

Longbottom, Will. 2020. 'Australia Bushfires: Sydney Suburb Is "Hottest Place on Earth" as Heat Creates Storms', *Sky News*, 5 January. Available at: https://news.sky.com/story/australia-bushfires-death-toll-rises-as-3000-army-reservists-are-deployed-11900782

Löwy, Michael. 2018. 'Why Ecosocialism: For a Red-Green Future' Great Transition Initiative: Towards Transformative Vision and Praxis', December. Available at: https://greattransition.org/publication/why-ecosocialism-red-green-future

Löwy, Michael. 2020. 'Thirteen Theses on the Imminent Ecological Catastrophe and the (Revolutionary) Means of Averting It', *International Viewpoint*, 4 February. Available at: http://internationalviewpoint.org/spip.php?article6391

Magdoff, Fred. 2018. 'Do Red and Green Mix?: A Roundtable', *Great Transition Initiative*, December. Available at: https://greattransition.org/roundtable/ecosocialism-fred-magdoff

Magdoff, Harry and Magdoff, Fred. 2005. 'Approaching Socialism', *Monthly Review*, 1 July. Available at: https://monthlyreview.org/2005/07/01/approaching-socialism/

Mair, Simon. 2018. 'Do Red and Green Mix? A contribution to an exchange on Why Ecosocialism: For a Red-Green Future'. https://greattransition.org/roundtable/ecosocialism-simon-mair

Maisuria, Alpesh. 2018. 'Mystification of Production and Feasibility of Alternatives: Social Class Inequality and Education', in Mike Cole (ed.), *Education, Equality and Human Rights: Issues of Gender, 'Race', Sexuality, Disability and Social Class*, 4th edition, London and New York: Routledge.

Majama, Koliwe. 2019. 'African Women Face Widening Technology Gap', *African School on Internet Governance*. Available at: https://afrisig.org/2019/04/01/african-women-face-widening-technology-gap/

Manchester, Julia. 2019. 'Andrew Yang Says "Fourth Industrial Revolution" Fueled Trump's Election', *The Hill*. Available at: https://thehill.com/hilltv/rising/435307-andrew-yang-says-trump-won-due-to-greatest-economic-and-technological

Mandel, Ernest. 1992. *Power and Money: A Marxist Theory of Bureaucracy*, London: Verso.

Marx, Karl. 1845. '*The German Ideology* Part I: Feuerbach. Opposition of the Materialist and Idealist Outlook A', *Idealism and Materialism*. Available at: www.marxists.org/archive/marx/works/1845/german-ideology/ch01a.htm#2

Marx, Karl. 1852. 'The Eighteenth Brumaire of Louis Bonaparte'. Available at: www.marxists.org/archive/marx/works/1852/18th-brumaire/ch01.htm

Marx, Karl. 1871. 'The Third Address'. May, [The Paris Commune] The Civil War in France Chapter 5. Available at: https://www.marxists.org/archive/marx/works/1871/civil-war-france/ch05.htm

Marx, Karl. 1887. *Capital Volume 1*, Moscow: Progress Publishers. Available at: https://oll.libertyfund.org/titles/marx-capital-a-critique-of-political-economy-volume-i-the-process-of-capitalist-production

Marx, Karl. 1893 [1967]. *Capital Volume 2*, Moscow: Progress Publishers. Available at: www.marxists.org/archive/marx/works/1885-c2/index.htm

Marx, Karl. 1894 [1966]. *Capital Volume 3*, Moscow: Progress Publishers. Available at: www.marxists.org/archive/marx/works/1894-c3/ch13.htm

Marx, Karl and Engels, Frederick. 1848 [2010]. 'Manifesto of the Communist Party Marxists Internet Archive (marxists.org)'. Available at: www.marxists.org/archive/marx/works/download/pdf/Manifesto.pdf

Mason, Rowena and Harvey, Fiona. 2020. 'Boris Johnson Doesn't Get Climate Change, Says Sacked COP 26 Chair', *The Guardian*, 4 February. Available at: www.theguardian.com/environment/2020/feb/04/sacked-cop-26-chair-claire-oneill-berates-boris-johnson-over-climate-record?CMP=Share_iOSApp_Other

McCarthy, Joe and Sanchez, Erica. 2019. '12 Female Climate Activists Who Are Saving the Planet', *Global Citizen*. Available at: www.globalcitizen.org/en/content/female-activists-saving-planet/

McGinnis, Devon. 2018. 'What Is the Fourth Industrial Revolution?', *Salesforce Blog*. Available at: www.salesforce.com/blog/2018/12/what-is-the-fourth-industrial-revolution-4IR.html

McGrath, Matt. 2019. 'Climate Change: COP25 Talks Open as "Point of No Return" in Sight', *BBC News*, 2 December. Available at: http://shamelnews.net/world/178607/Climate-change-COP25-talks-to-open-as-point-of-no-return-in-sight?source=true

McGrath, Matt. 2020a. 'Climate Change: Australia Fires Will Be "Normal" in Warmer World', *BBC News*, 14 January. Available at: www.bbc.co.uk/news/science-environment-51094919

McGrath, Matt. 2020b. 'Climate Change: Last Decade Confirmed as Warmest on Record', *BBC News*, 15 January. Available at: www.bbc.co.uk/news/science-environment-51111176

McGregor, Helen, Gergis, Joelle, Abram, Nerilie and Phipps, Steven. 2016. 'The Industrial Revolution Kick-started Global Warming Much Earlier than We Realised', *The Conversation*, 24 August. Available at: https://theconversation.com/the-industrial-revolution-kick-started-global-warming-much-earlier-than-we-realised-64301

McLaren, Peter. 2000. *Che Guevara, Paulo Freire and the Pedagogy of Revolution*, Oxford: Rowman and Littlefield.

McMurtry, John. 2000. 'Education, Struggle and the Left Today', *International Journal of Educational Reform* 10(2).

McQue, Katie, Townsend, Mark and Armour, Katie. 2019. 'Windrush Scandal Continues as Chagos Islanders Are Pressed to "Go Back"', *The Guardian*, 28 July. Available at: www.theguardian.com/world/2019/jul/28/windrush-scandal-continues-in-crawley-as-chagos-islanders-told-go-back

Mecklin, John. 2020. 'It Is 100 Seconds to Midnight', *Bulletin of the Atomic Scientists*. Available at: https://thebulletin.org/doomsday-clock/current-time/

Mellor, Mary. 1997. *Feminism & Ecology*. New York: New York Univerity Press.

Mellor, Mary. 2018. 'Do Red and Green Mix?: A Roundtable', *Great Transition Initiative*, December. Available at: www.greattransition.org/roundtable/ecosocialism-mary-mellor

Mellor, Mary. 2019. 'An Eco-feminist Proposal', *New Left Review*, 116/117, March/June.

Merrick, Rob. 2020. 'Boris Johnson Pledges to Ban New Petrol and Diesel Cars from 2035 After Claims His Climate Plans Are in "Chaos"', *The Independent*, 4 February. Available at: www.independent.co.uk/news/uk/politics/uk-ban-petrol-diesel-cars-boris-johnson-government-a9315631.html?amp

Millett, Briana. 2020. 'In Full: Greta Thunberg's Speech at Bristol Climate Strike', *BristolLive*, 28 February. Available at: www.bristolpost.co.uk/news/bristol-news/greta-thunberg-speech-in-full-3897406

Molyneux, John. 2019. 'Extinction Rebellion: A Socialist Perspective', *Links International Journal of Socialist Renewal*, 4 November. Available at: http://links.org.au/extinction-rebellion-socialist-perspective

Moore, Jason W. 2017. 'The Capitalocene Part I: On the Nature & Origins of Our Ecological Crisis', *Journal of Peasant Studies* 44(3), April, 594–630.

National Centers for Environmental Information. 2019. 'Global Climate Report – January 2019'. Available at: www.ncdc.noaa.gov/sotc/global/201901

National Centers for Environmental Information. 2020. 'Assessing the Global Climate in January 2020. January 2020 Was the Warmest January on Record for the Globe'. Available at: www.ncei.noaa.gov/news/global-climate-202001

National Defense Strategy. 2018. 'Summary of the National Defense Strategy Sharpening the American Military's Competitive Edge'. Available at: https://dod.defense.gov/Portals/1/Documents/pubs/2018-National-Defense-Strategy-Summary.pdf

Nicks, Denver. 2017. 'The Craziest Senator', *The Daily Beast*, 14 July. Available at: www.thedailybeast.com/the-craziest-senator

Ni Loideain, Nóra and Adams, Rachel. 2019. 'From Alexa to Siri and the GDPR: The Gendering of Virtual Personal Assistants and the Role of EU Data Protection Law'. Available at: https://staff.ki.se/sites/default/files/migrate/2019/04/16/ni_loideain_and_adams_siri_alexa_and_gdpr_march_2019.pdf

North, David and Kishore, Joseph. 2020. 'The Decade of Socialist Revolution Begins', *World Socialist Web Site (WSWS)*, 3 January. Available at: www.wsws.org/en/articles/2020/01/03/pers-j03.html

Office of Naval Research. Undated. 'Directed Energy Weapons: Counter Directed Energy Weapons and High Energy Lasers'. Available at: www.onr.navy.mil/en/Science-Technology/Departments/Code-35/All-Programs/aerospace-science-research-351/directed-energy-weapons-cdew-and-high-energy-lasers

O'Neill, Deirdre and Wayne, Mike. 2017. 'On Intellectuals', *Historical Materialism*, 8 October. Available at: www.historicalmaterialism.org/blog/intellectuals

Oxley, Greg. 2001. 'The Paris Commune of 1871', *In Defence of Marxism*, 16 May. Available at: www.marxist.com/paris-commune-of-1871.htm

Peters, Andrea. 2019. 'A Magnificent Account of Stalin's Opponents in the USSR', *World Socialist Web Site (WSWS)*, 21 December. Available at: www.wsws.org/en/articles/2019/12/21/rogo-d21.html

Phillips, Tom. 2019a. 'Chaos, Chaos, Chaos': A Journey Through Bolsonaro's Amazon', *The Guardian*, 9 September. Available at: www.theguardian.com/environment/2019/sep/09/amazon-fires-brazil-rainforest

Phillips, Tom. 2019b. 'Greta Thunberg Labelled a "Brat" by Brazil's Far-right Leader Jair Bolsonaro', *The Guardian*, 10 December. Available at: www.theguardian.com/environment/2019/dec/10/greta-thunberg-jair-bolsonaro-brazil-indigenous-amazon

Preston, John. 2019. *Grenfell Tower: Preparedness, Race and Disaster Capitalism*, London: Palgrave Pivot.

Qasim, Arsheen. 2019. '5 Ways Climate Change Affects Women and Girls', *ActionAid*, 29 November. Available at: www.actionaid.org.uk/blog/voices/2019/11/29/5-ways-climate-change-affects-women-and-girls?gclid=EAIaIQobChMIs4GOmcyK6AIVWeDtCh3VYQHwEAAYASAAEgI0U_D_BwE

Red Pepper. 2019. 'An Open Letter to Extinction Rebellion', 3 May. Available at: www.redpepper.org.uk/an-open-letter-to-extinction-rebellion/

Reed, Kevin. 2020a. 'Clearview AI Facial Recognition Tool Being Used by More than 600 US Police Agencies', *World Socialist Web Site (WSWS)*, 22 January. Available at: www.wsws.org/en/articles/2020/01/22/face-j22.html

Reed, Kevin. 2020b. 'Amazon Employees Defy Management and Publicly Protest Corporate Policies', *World Socialist Web Site (WSWS)*, 30 January. Available at: www.wsws.org/en/articles/2020/01/30/amzn-j30.html

Rehman, Asad. 2019. 'The Global Elite Is Destroying Our Planet. So Why Are Extinction Rebellion Activists the Ones in the Dock?', *The Independent*, 18 April. Available at: www.independent.co.uk/voices/extinction-rebellion-climate-change-global-elite-capitalism-a8876711.html

Reuters. 2019. 'Victoria Falls Dries to a Trickle After Worst Drought in a Century', *The Guardian*, 7 December. Available at: https://amp.theguardian.com/world/2019/dec/07/victoria-falls-dries-to-a-trickle-after-worst-drought-in-a-century

Richter, Stephan and Bott, Uwe. 2019. 'Davos Has Learned to Fake Populism', *Foreign Policy*, 24 January. Available at: https://foreignpolicy.com/2019/01/24/davos-has-learned-to-fake-populism/

Rikowski, Glenn and Ocampo Gonzalez, Aldo. 2018. 'Interview on Marxism, Critical Pedagogy and Inclusive Education: Discussions for a Revolutionary Discourse', *(Glenn Rikowski interviewed by Aldo Ocampo Gonzalez), the Center for Latin American Studies on Inclusive Education (CELEI)*, March. Available at: www.celei.cl/wp-content/uploads/2018/03/Entrevista-sobre-Marximo-Pedagog%C3%ADa-Cr%C3%ADtica-y-Educaci%C3%B3n-Inclusiva_Dr.-Glenn-Rikowski_UK.pdf

Rogovin, Vadim Z. 2019. *Bolsheviks Against Stalinism, 1928–1933: Leon Trotsky and the Left Opposition*, Mehring Books: Oak Park, MI.

Saltman, Kenneth J. 2007. 'Schooling in Disaster Capitalism: How the Political Right Is Using Disaster to Privatize Public Schooling', *Teacher Education Quarterly* 34(2).

Sandlin, Jennifer A., O'Malley, Michael P. and Burdick, Jake. 2011. 'Mapping the Complexity of Public Pedagogy Scholarship: 1894–2010', *Review of Educational Research*, September, pp. 338–375.

Sandlin, Jennifer A., Schultz, Brian D. and Burdick, Jake. 2010. *Handbook of Public Pedagogy*, New York: Routledge.

Schellnhuber, Hans. 2018. '2018 Aurelio Peccei Lecture: Climate, Complexity, Conversion', *50th Anniversary Summit of the Club of Rome*, 17 October, Augustinian Patristic Institute, Rome, Italy. Available at: www.youtube.com/embed/Rtg5QJlb484

Scholz, Trebor. 2015. 'Think Outside the Boss: Cooperate Alternatives to the Sharing Economy', *the 8th Annual Eric N. Schocket Memorial Lecture on Class and Culture*, 5 April, Public Seminar. Available at: http://publicseminar.org/2015/04/think-outside-the-boss/

Schubert, William H. 1986. *Curriculum: Perspective, Paradigm, and Possibility*, New York: Macmillan.

Schuppe, Jon. 2019. 'Amazon Is Developing High-tech Surveillance Tools for an Eager Customer: America's Police', *NCB News*, 8 August. Available at: www.nbcnews.com/tech/security/amazon-developing-high-tech-surveillance-tools-eager-customer-america-s-n1038426

Schwab, Klaus. 2015. 'The Fourth Industrial Revolution: What It Means and How to Respond', *Foreign Affairs*, 12 December. Available at: www.foreignaffairs.com/articles/2015-12-12/fourth-industrial-revolution

Schwab, Klaus. 2016. *The Fourth Industrial Revolution*, Cologny and Geneva: World Economic Forum.

Schwab, Klaus. 2018. 'Globalization 4.0 – What Does It Mean?', *World Economic Forum*. Available at: www.weforum.org/agenda/2018/11/globalization-4-what-does-it-mean-how-it-will-benefit-everyone/

Schwab, Klaus. 2019. 'What Kind of Capitalism Do We Want?', *Project Syndicate*, 2 December. Available at: www.project-syndicate.org/commentary/stakeholder-capitalism-new-metrics-by-klaus-schwab-2019-11

Scott, Martin. 2019. 'Government Has No Solution to Australia's Water Crisis', *World Socialist Web Site (WSWS)*, 16 October. Available at: www.wsws.org/en/articles/2019/10/16/wate-o16.html

Scoville, Heather. 2018. 'Thought.com', 5 July. Available at: www.thoughtco.com/the-5-major-mass-extinctions-4018102

Sengupta, Somini. 2019. 'Climate Protesters and World Leaders: Same Planet, Different Worlds', *The New York Times*, 21 September. Available at: www.nytimes.com/2019/09/21/climate/united-nations-climate-change.html

Serge, Victor. 1930 [1972]. *Year One of the Russian Revolution*, New York: Holt, Reinhart, and Winston. Victor Serge Internet Archive (marxists.org) 2005. Translation, editor's Introduction, and notes 1972 by Peter Sedgwick. Available at: www.marxists.org/archive/serge/1930/year-one/index.htm

Shapiro, Ben. 2016. 'Why Trump Fans Keep Using the Slur "Globalist"', *The Daily Wire*, 2 August. Available at: www.dailywire.com/news/why-trump-fans-keep-using-slur-globalist-ben-shapiro

Shattuck, John, Watson, Amanda and McDole, Matthew. 2018. *Trump's First Year: How Resilient Is Liberal Democracy in the US?* Cambridge, MA: Carr Center for Human Rights Policy Harvard Kennedy School. Available at: https://carrcenter.hks.harvard.edu/files/cchr/files/trumpsfirstyeardiscussionpaper.pdf

Simon, Roger. 1995. 'Broadening the Vision of University-based Study of Education: The Contribution of Cultural Studies', *The Review of Education/Pedagogy/Cultural Studies* 12(1).

Sky News. 2019. 'Australian Bushfires to Get Worse as Strong Winds and High Temperatures Create "Catastrophic" Conditions', *Sky News*, 21 December. Available at: https://news.sky.com/story/oz-pm-returns-home-as-bushfires-create-catastrophic-conditions-11891891

Smith, R. Jeffrey. 2019. 'Hypersonic Missiles Are Unstoppable. And They're Starting a New Global Arms Race', *The New York Times Magazine*, 19 June. Available at: www.nytimes.com/2019/06/19/magazine/hypersonic-missiles.html.

Squirrell, Tim. 2017. 'Linguistic Data Analysis of 3 Billion Reddit Comments Shows the Alt-right Is Getting Stronger', *Quartz*, 18 August. Available at: https://qz.com/1056319/what-is-the-alt-right-a-linguistic-data-analysis-of-3-billion-reddit-commentsshows-a-disparate-group-that-is-quickly-uniting/

Street, Paul. 2019. 'The Woman-Hater-in-Chief', *Counterpunch*, 16 December. Available at: www.counterpunch.org/2019/12/16/the-woman-hater-in-chief/

Stylianou, Nassos, Guibourg, Clara, Dunford, Daniel and Rodgers, Lucy. 2019. 'Climate Change: Where We Are in Seven Charts and What You Can Do to Help', *BBC News*, 18 April. Available at: www.bbc.co.uk/news/science-environment-46384067

Suarez-Villa, Luis. 2009. *Technocapitalism: A Critical Perspective on Technological Innovation and Corporatism*, Philadelphia: Temple University Press.

Surin, Kenneth. 2019. 'Australia's Big Smoke', *Counterpunch*, 11 December. Available at: www.counterpunch.org/2019/12/11/australias-big-smoke/

Taaffe, Peter. 2014. A 'third industrial revolution'. *Socialism Today*, 183. Available at: http://socialismtoday.org/archive/183/technology.html

Tanuro, Daniel. 2019 'Hands off Greta Thunberg', *Climate and Capitalism*, 10 October. Available at: https://climateandcapitalism.com/2019/10/10/hands-off-greta-thunberg/

Taylor, Adam. 2019. 'Australia's Prime Minister Pledges to Outlaw Climate Boycotts, Arguing They Threaten the Economy', *The Washington Post*, 1 November. Available at: www.washingtonpost.com/world/2019/11/01/australias-prime-minister-pledges-outlaw-climate-boycotts-arguing-they-threaten-economy/

Taylor, Frederick Winslow. 1911 [2003]. *The Principles of Scientific Management*, Mineola, NY: Dover Publications Inc.

Taylor, Matthew. 2019 'What Are Extinction Rebellion's Key Demands?', *The Guardian*, 8 October. Available at: www.theguardian.com/environment/2019/oct/08/what-are-extinction-rebellion-key-demands-climate-emergency

The Indigenous Environmental Network (IEN). 2019. 'Talking Points on the AOC-Markey Green New Deal (GND) Resolution'. Available at: https://www.ienearth.org/talking-points-on-the-aoc-markey-green-new-deal-gnd-resolution/

The Spark. 1977. 'The Paris Commune, 1871', 31 December. Available at: https://the-spark.net/o_pariscom.html

Thomas, Peter D. 2009. *The Gramscian Moment: Philosophy, Hegemony and Marxism: Historical Materialism*, Volume 24, Leiden, The Netherlands: Brill.

Thompson, Clive. 2019 'How 19th Century Scientists Predicted Global Warming', *JSTOR Daily*, 17 December. Available at: https://daily.jstor.org/how-19th-century-scientists-predicted-global-warming/

Treanor, Jill. 2016. 'Fourth Industrial Revolution Set to Benefit Richest, UBS Report Says', *The Guardian*, 19 January. Available at: www.theguardian.com/business/2016/jan/19/fourth-industrial-revolution-set-to-benefit-richest-ubs-report-says

Turner, Terisa E., and Brownhill, Leigh S. 2001a. '"Women Never Surrendered": The Mau Mau and Globalization from Below in Kenya, 1980–2000.' In *There is an Alternative: Subsistence and Worldwide Resistance to Corporate Globalization*, edited by Bennholdt-Thomsen, V., Faraclas, Nicholas, and Von Werlhof, Claudia London: Zed Books.

Turner, Terisa E., and Brownhill, Leigh S. 2001b. 'African Jubilee: Mau Mau Resurgence and the Fight for Fertility in Kenya, 1986–2002.' *Canadian Journal of Development Studies* XXII: 1037–88.

United Nations. 1987. 'Report of the World Commission on Environment and Development: Our Common Future'. Available at: www.un-documents.net/wced-ocf.htm

United Nations. 1992. 'Agenda 21'. Available at: https://sustainabledevelopment.un.org/outcomedocuments/agenda21

United Nations Climate Change. 2020. 'Conference of the Parties (COP)'. Available at: https://unfccc.int/process/bodies/supreme-bodies/conference-of-the-parties-cop

Van Auken, Bill. 2020. 'US deploys "usable" nuclear weapon amid continuing war threats against Iran'. World Socialist Web Site (WSWS), 1 February. Available at: https://www.wsws.org/en/articles/2020/02/01/pers-f01.html

Vaughan, Adam. 2019. 'IPCC Report: Sea Levels Could Be a Metre Higher by 2100', *New Scientist*. Available at: www.newscientist.com/article/2217611-ipcc-report-sea-levels-could-be-a-metre-higher-by-2100/#ixzz67oXOBsxl

VOA. 2019. 'With Trump Out, Davos Chief Eyes Fixing World Architecture', 20 January. Available at: www.voanews.com/europe/trump-out-davos-chief-eyes-fixing-world-architecture

Waheed, Rida, Sarwar, Sahar and Wei, Chen. 2019. 'The Survey of Economic Growth, Energy Consumption and Carbon Emission', *Energy Reports*, 5 November. Available at: www.sciencedirect.com/science/article/pii/S2352484719302082

Wainwright, Oliver. 2012 'Billionaires' Basements: The Luxury Bunkers Making Holes in London Streets', *The Guardian*, 9 November. Available at: www.theguardian.com/artanddesign/2012/nov/09/billionaires-basements-london-houses-architecture

Walker, Peter. 2019. 'MPs Endorse Corbyn's Call to Declare Climate Emergency', *The Guardian*, 1 May. Available at: www.theguardian.com/environment/2019/may/01/declare-formal-climate-emergency-before-its-too-late-corbyn-warns

Watts, Jonathan. 2018. 'We Have 12 Years to Limit Climate Change Catastrophe, Warns UN', *The Guardian*. 8 October. Available at: www.theguardian.com/environment/2018/oct/08/global-warming-must-not-exceed-15c-warns-landmark-un-report

Webber, Frances. 2019. 'Johnson's Immigration Policies: Hostile Chaos?', *Institute of Race Relations*, 30 August. Available at: www.irr.org.uk/news/johnsons-immigration-policies-hostile-chaos/

Weedon, Alan. 2019. 'What Exactly Is a Railgun, and Do We Need to Start Building Bunkers Again?', *ABC News*, 4 January. Available at: www.abc.net.au/news/2019-01-03/the-railgun-explained-and-what-it-means-for-us-china-relations/10683526

Whiting, Kate. 2020. '5 Shocking Facts About Inequality, According to Oxfam's Latest Report', *World Economic Forum*, 20 January. Available at: www.weforum.org/agenda/2020/01/5-shocking-facts-about-inequality-according-to-oxfam-s-latest-report/

Wilson, Jason. 2015. '"Cultural Marxism": A Uniting Theory for Rightwingers Who Love to Play the Victim', *The Guardian*, 19 January. Available at: www.theguardian.com/commentisfree/2015/jan/19/cultural-marxism-a-uniting-theory-for-right wingers-who-love-to-play-the-victim

Wilson, Jason. 2019. 'Eco-fascism Is Undergoing a Revival in the Fetid Culture of the Extreme Right', *The Guardian*, 20 March. Available at: www.theguardian.com/world/commentisfree/2019/mar/20/eco-fascism-is-undergoing-a-revival-in-the-fetid-culture-of-the-extreme-right

Wood, Vincent. 2019. 'Greta Thunberg Responds to Trump's Attack After Her Time Person of the Year Win', *The Independent*, 13 December. Available at: www.independent.co.uk/news/world/europe/greta-thunberg-trump-tweet-time-person-of-the-year-bio-twitter-a9243961.html

World Economic Forum (WEF). 2019. 'Fourth Industrial Revolution for the Earth'. Available at: www.weforum.org/projects/fourth-industrial-revolution-and-environment-the-stanford-dialogues

World Population Review. 2020a. '2020 World Population by Country'. Available at: http://worldpopulationreview.com/

World Population Review. 2020b. 'Continent and Region Populations 2020'. Available at: https://worldpopulationreview.com/continents/

Yan, Nicolas. 2016. 'Automated Inequality'. *Harvard Political Review*, 2 October. | https://harvardpolitics.com/world/automation/

Index

Printed in the United States
by Baker & Taylor Publisher Services